Cities and Climate Change

Climate change is one of the most significant global challenges facing the world today. It is also a critical issue for the world's cities. Now home to over half the world's population, urban areas are significant sources of greenhouse-gas emissions and are vulnerable to the impacts of climate change.

Responding to climate change is a profound challenge. A variety of actors are involved in urban climate governance, with municipal governments, international organizations and funding bodies pointing to cities as key arenas for response. This book provides the first critical introduction to these challenges, giving an overview of the science and policy of climate change at the global level and the emergence of climate change as an urban policy issue. It considers the challenges of governing climate change in the city in the context of the changing nature of urban politics, economics, society and infrastructures. It looks at how responses for mitigation and adaptation have emerged within the city, and the implications of climate change for social and environmental justice.

Drawing on examples from cities in the north and south, and richly illustrated with detailed case studies, this book will enable students to understand the potential and limits of addressing climate change at the urban level and to explore the consequences for our future cities. It will be essential reading for undergraduate students across the disciplines of geography, politics, sociology, urban studies, planning, and science and technology studies.

Harriet Bulkeley is Professor of Geography, Energy and Environment in the Department of Geography, and Deputy Director of the Durham Energy Institute, Durham University, UK. Her research interests are in the nature and politics of environmental governance and focus on climate change and urban sustainability.

Routledge critical introductions to urbanism and the city

Edited by Malcolm Miles, University of Plymouth, UK,
and John Rennie Short, University of Maryland, USA

International Advisory Board:

Franco Bianchini
Kim Dovey
Stephen Graham
Tim Hall
Phil Hubbard
Peter Marcuse

Jane Rendell
Saskia Sassen
David Sibley
Erik Swyngedouw
Elizabeth Wilson

The series is designed to allow undergraduate readers to make sense of, and find a critical way into, urbanism. It will:

- cover a broad range of themes
- introduce key ideas and sources
- allow the author to articulate her/his own position
- introduce complex arguments clearly and accessibly
- bridge disciplines, and theory and practice
- be affordable and well designed.

The series covers social, political, economic, cultural and spatial concerns. It will appeal to students in architecture, cultural studies, geography, popular culture, sociology, urban studies and urban planning. It will be trans-disciplinary. Firmly situated in the present, it also introduces material from the cities of modernity and post-modernity.

Published:

Cities and Consumption – Mark Jayne
Cities and Cultures – Malcolm Miles
Cities and Nature – Lisa Benton-Short and John Rennie Short
Cities and Economies – Yeong-Hyun Kim and John Rennie Short
Cities and Cinema – Barbara Mennel
Cities and Gender – Helen Jarvis with Paula Kantor and Jonathan Cloke
Cities and Design – Paul L. Knox
Cities, Politics and Power – Simon Parker
Cities and Sexualities – Phil Hubbard
Children, Youth and the City – Kathrin Hörshelmann and Lorraine van Blerk
Cities and Photography – Jane Tormey
Cities and Climate Change – Harriet Bulkeley

Forthcoming:

Cities, Risk and Disaster – Christine Wamsler

Cities and Climate Change

Harriet Bulkeley

LONDON AND NEW YORK

First published 2013
by Routledge
2 Park Square, Milton Park, Abingdon, Oxon OX14 4RN

Simultaneously published in the USA and Canada
by Routledge
711 Third Avenue, New York, NY 10017

Routledge is an imprint of the Taylor & Francis Group, an informa business

© 2013 Harriet Bulkeley

The right of Harriet Bulkeley to be identified as author of this work has been asserted by her in accordance with sections 77 and 78 of the Copyright, Designs and Patents Act 1988.

All rights reserved. No part of this book may be reprinted or reproduced or utilised in any form or by any electronic, mechanical, or other means, now known or hereafter invented, including photocopying and recording, or in any information storage or retrieval system, without permission in writing from the publishers.

Trademark notice: Product or corporate names may be trademarks or registered trademarks, and are used only for identification and explanation without intent to infringe.

British Library Cataloguing in Publication Data
A catalogue record for this book is available from the British Library

Library of Congress Cataloging in Publication Data
A catalog record for this book has been requested

ISBN: 978-0-415-59704-3 (hbk)
ISBN: 978-0-415-59705-0 (pbk)
ISBN: 978-0-203-07720-7 (ebk)

Typeset in Times New Roman
by Florence Production Ltd, Stoodleigh, Devon

Printed and bound by CPI Group (UK) Ltd, Croydon, CR0 4YY

Contents

List of Figures viii

List of Tables ix

List of Boxes x

List of Case studies xi

Acknowledgements xii

1 Climate change – an urban problem? 1

Introducing climate change 1

What has climate change got to do with the city? 5

Urban worlds and climate-change futures 9

Chapter outline 14

Discussion points 16

Further reading and resources 16

2 Climate risk and vulnerability in the city 18

Introduction 18

The urban impacts of climate change 20

Climate change and urban vulnerability 28

Conclusions 42

Discussion points 43

Further reading and resources 43

3 Accounting for urban GHG emissions — 45

Introduction 45

Assessing the urban contribution to climate change 47

Urban difference and the drivers of GHG emissions 58

Conclusions 69

Discussion points 69

Further reading and resources 70

4 Governing climate change in the city — 71

Introduction 71

Charting the emergence of urban climate-change responses 74

Understanding the nature of urban climate governance 91

Conclusions 104

Discussion points 105

Further reading and resources 105

5 Climate-change mitigation and low-carbon cities — 106

Introduction 106

What is mitigation? 108

Making mitigation policy 109

Municipal climate-change mitigation in action 118

The drivers and challenges of mitigation 132

Conclusions 138

Discussion points 140

Further reading and resources 141

6 Urban adaptation – towards climate-resilient cities?   **142**

Introduction   142

What is adaptation?   144

Making adaptation policy   150

Municipal climate-change adaptation in practice   163

Drivers and barriers for urban climate-change adaptation   179

Conclusions   187

Discussion points   188

Further reading and resources   188

7 Climate-change experiments and alternatives in the city   **190**

Introduction   190

Sowing the seeds of change?   192

Experiments in practice   200

Alternatives: moving beyond the mainstream?   215

Conclusions   226

Discussion points   227

Further reading and resources   227

8 Conclusions   **229**

Changing climate, changing cities?   230

Urban futures: towards climate justice?   235

Bibliography   238

Index   255

Figures

1.1	Imagining urban climate-change futures	13
2.1	Map showing cities predicted to have populations most exposed to sea-level rise in the 2070s under conditions of climate change	30
2.2	Map of the cities predicted to have assets most exposed to sea-level rise in the 2070s under conditions of climate change	31
2.3	Storm clouds over Cape Town	38
3.1	ClimateCam billboard, Newcastle	51
3.2	Amman City	56
3.3	Daily dairy moments	68
4.1	C40 Member and affiliate cities 2011	81
4.2	Melbourne: climate governance in action	90
5.1	The City of São Paulo	115
5.2	Solar hot-water systems in São Paulo's social-housing projects	116
5.3	Existing air-conditioning systems and norms about indoor thermal comfort in Hong Kong may limit the potential for energy efficiency	139
6.1	Residents of rowhomes in Philadelphia may be particularly vulnerable to increased incidents of heatwaves	157
6.2	Climate adaptation in Ga Mashie, Ghana	162
6.3	Concrete recharge pits and bore wells in Mumbai, designed using the principles of Rajasthan stepwells	175
7.1	Distribution of experiments in different sectors in the UTACC database	199
7.2	The frequency with which actors participate in climate-change experiments in the UTACC database	200
7.3	Advertisement for PowerSmart at the Star Ferry Terminal, Hong Kong	213
7.4	Community gardens in Brixton, London	221

Tables

1.1	Predicted regional impacts of climate change	3
1.2	Cities and climate change: part of the problem, part of the solution?	8
2.1	Climate risks and impacts for cities	21
2.2	Climate impacts and the interactions between energy and other urban systems	33
2.3	Climate risks and urban children	41
3.1	Outline for a programme of activities for Amman	57
3.2	Which are the top ten? Three different ways of evaluating the global contribution of cities	60
4.1	The emergence of climate-change governance in Melbourne	88
4.2	Modes of governing climate change in the city	94
4.3	Drivers and barriers for urban climate-change responses	101
5.1	Drivers and barriers for climate-change mitigation	133
6.1	Levels of adaptation	149
6.2	Drivers and barriers for urban climate-change adaptation	181
7.1	Examples of climate-change experiments in the Australian water sector	208
7.2	Challenges facing the mainstreaming of experimentation	214

Boxes

1.1	What are GHG emissions?	2
1.2	Virtual urban responses to climate change: LogiCity	12
2.1	Defining climate-change vulnerability	19
2.2	Cities at risk	22
2.3	The urban heat island effect	26
2.4	The social impacts of floods in urban communities	39
3.1	Principles for the design of the International Local Government GHG Emissions Analysis Protocol	48
3.2	Defining Scope I, II and III GHG emissions	49
3.3	ClimateCam, Newcastle, Australia	50
3.4	What are Toronto's per capita emissions?	62
4.1	The new governance?	73
4.2	ICLEI South Asia	79
4.3	Manchester is My Planet: mobilizing the community?	99
4.4	Policy entrepreneurs	103
5.1	Milestone-based approaches to climate mitigation at the urban level: CCP methodology	111
5.2	Thane's solar hospital	124
5.3	Smart cities: low-carbon future?	131
6.1	What is adaptation?	145
6.2	Defining resilience	147
6.3	Informal adaptation in Accra	161
6.4	Toronto's heat health alert system	166
6.5	Community-based climate-resilient development in Durban	172
6.6	Rainwater harvesting in Mumbai	174
6.7	Integrating climate-change adaptation into infrastructure planning in New York	177
7.1	Urban emissions trading experiments	203
7.2	About the Solar American Cities programme	210
7.3	Oxford is My World	211

Case studies

2.1	Climate change and vulnerability in Cape Town	36
3.1	Calculating carbon finance for Amman	54
4.1	Governing climate change in Melbourne	85
5.1	Climate-change mitigation in São Paulo	113
6.1	Adapting to climate change in Philadelphia	154
7.1	Climate experiments and alternatives in London	219

Acknowledgements

Although this book appears to be the work of a single author, in reality it has been a product of a great many people, only some of whom I can explicitly acknowledge here. First, my thanks to the editorial team at Routledge for bearing with me over the production process, and particularly to Faye Leerink and Andrew Mould for their patience. Second, a number of colleagues have helped me to develop my understanding of urban responses to climate change over the years, and I would particularly like to thank Michelle Betsill, JoAnn Carmin, Vanesa Castán Broto, Gareth Edwards, Simon Guy, Mike Hodson, Matthew Hoffmann, Kristine Kern, Simon Marvin and Heike Schroeder for their engagement with my work. My research and this book project in particular have also been considerably advanced through the support of an ESRC Climate Change Fellowship – Urban Transitions: climate change, global cities and the transformation of sociotechnical networks (award number: RES-066-27-0002).

This book has itself benefited beyond all measure by the input of a great team of graduate students and research associates. I would like to thank Gerald Aiken, Catherine Button, Vanesa Castán Broto, Gareth Edwards, Sara Fuller, Andrés Luque and Jonathan Silver for their direct contributions to the book, including photographs, text boxes and case studies. Andrea Armstrong did an amazing job of helping me to pull the text, figures and references together, and I am very grateful for her skilled editorial work. I would also like to thank the City of Newcastle, Australia, for its kind permission to use the photograph in Figure 3.1, Siegfried Atteneder for permission to use his photograph of Amman City (Figure 3.2), and WWF and Imperial College London for allowing me to reproduce the diagram used in Figure 3.3. The cartography unit at the Department of Geography, Durham University, UK, provided Figure 4.1, for which many thanks.

Last, but by no means least, my gratitude extends to my partner, Pete, and my children, Elodie and Théa, who have borne much of the burden of having this book in the house for so long. Without them, none of this would be possible.

1 Climate change – an urban problem?

Introducing climate change

Over the past two decades, the term 'climate change' has for many acquired a well-worn familiarity. Whether on the pages of the national newspapers, through events such as the NDTV–Toyota's Greenathon in India, the WWF Earth Hour campaign, or the commitments of an ever-growing number of businesses to address the issue, climate change has entered into the popular vocabulary (Hulme 2009). Yet, at the same time, climate change remains a distant concern. In many parts of the world, and for many people, the extent and consequences of the ways in which human activities may be leading to long-term changes in the global atmosphere are obscured. This may be because of continued scientific controversy and political attempts to deny the significance or importance of the issue, or it may be because other, pressing, concerns dominate daily life. Elsewhere, climate change has a distant presence; it is an issue that rumbles on in the background to daily lives, surfacing at particular moments as, for example, international leaders seek agreement at the latest negotiations or another climate-related disaster strikes.

These qualities have made climate change one of the most taken-for-granted and most debated issues of the contemporary era. Although most agree that it is a serious issue that requires sustained attention and action, over two decades of international negotiation have yet to deliver a significant global response. The longevity of these debates, and the lack of resolution at the international level, could lead many to dismiss climate change as a lost cause. Given this context, it is worth reminding ourselves just what might be at stake. In their 2007 Fourth Assessment Report, the Intergovernmental Panel on Climate Change (IPCC) found that the weight of scientific evidence indicated that, 'warming of the climate system is unequivocal, as is now evident from observations of increases in global average air and ocean temperatures, widespread melting of snow and ice and rising global average sea level', and that this was beginning to have an effect regionally on natural systems (IPCC 2007a). The IPCC found that, over the period from 1970 to 2004, there had been a 70 per cent increase in global emissions of greenhouse gases (GHGs) (Box 1.1) and concluded that, 'most of the observed increase in global average temperatures since the mid-20th century is very likely due to the observed increase in anthropogenic GHG concentrations' (IPCC 2007c). As a result of changing climatic conditions, the IPCC identified a series of potential impacts on different human and natural systems in different regions of the world, illustrated in Table 1.1.

Although the IPCC's language and its cautious calculation of the potential for global warming and the ensuing impacts have become an accepted part of society's understanding of climate change, for others it is precisely this language that has served to diminish society's engagement with the potentially far-reaching implications of a changing climate. From this perspective, what John Urry has termed the 'new catastrophism', climate change, is regarded as a symptom of the failures of industrial society, and, as such, 'there are no guarantees that the increasing prosperity, wealth, movement and connectivity of the industrial period will continue forever. A new and darker epoch may

---

BOX 1.1

What are GHG emissions?

Greenhouse gases are those gaseous constituents of the atmosphere, both natural and anthropogenic, that absorb and emit radiation at specific wavelengths within the spectrum of infrared radiation emitted by the Earth's surface, the atmosphere and clouds. This property causes the greenhouse effect. Water vapour (H_2O), carbon dioxide (CO_2), nitrous oxide (N_2O), methane (CH_4) and ozone (O_3) are the primary GHGs in the Earth's atmosphere. Moreover, there are a number of entirely human-made GHGs in the atmosphere, such as the halocarbons and other chlorine- and bromine-containing substances, dealt with under the Montreal Protocol. Besides carbon dioxide, nitrous oxide and methane, the Kyoto Protocol deals with the GHGs sulphur hexafluoride, hydrofluorocarbons and perfluorocarbons.

(IPCC 2007b)

Table 1.1 Predicted regional impacts of climate change

Region	Potential impacts of climate change
Africa	By 2020, between 75 and 250 million people may be exposed to increased water stress.
	Towards the end of the 21st century, projected sea-level rise will affect low-lying coastal areas with large populations.
Asia	By the 2050s, fresh-water availability, particularly in large river basins, is predicted to decrease.
	Coastal areas, especially heavily populated mega-delta regions, will be at risk from increased flooding.
	Climate change is projected to compound the pressures on natural resources associated with rapid urbanization and economic development.
Australia and New Zealand	By 2020, significant loss of biodiversity is projected to occur, including in the Great Barrier Reef and Queensland Wet Tropics.
	By 2030, water-security problems are projected to intensify in southern and eastern Australia.
	Coastal development and population growth will exacerbate risks from sea-level rise, storms and coastal flooding by 2050.
Europe	In southern Europe, climate change is projected to exacerbate high temperatures and drought, and reduce water availability.
	Climate change is also projected to increase the number of heatwaves and the frequency of wildfires.
Latin America	There is a risk of significant biodiversity loss through species extinction in many areas of tropical Latin America.
	Changes in precipitation patterns and the disappearance of glaciers are projected to significantly affect water availability.
North America	Warming in western mountains is projected to exacerbate competition for over-allocated water resources.
	Urban heatwaves are expected to increase in number, intensity and duration, with potential for adverse health impacts.
	Coastal communities and habitats will be increasingly stressed by climate-change impacts interacting with development and pollution.

Source: Adapted from IPCC 2007c

lie in the quite near future' (Urry 2011: 36). Critical thinkers such as Bill McKibben, James Lovelock and Mark Lynas have argued that the increased levels of GHG emissions in the atmosphere are likely to lead to a range of significant and catastrophic effects that would disrupt global markets, create flows of climate refugees, and challenge existing forms of political order (McKibben 2011; Lovelock 2009; Lynas 2004). In this sense, climate change is not a problem that can be managed through the application of new scientific knowledge and policy instruments, but rather requires fundamental shifts in the ways in which economies and societies are organized and operate.

Although they differ considerably in how the challenge of climate change is understood, both the 'scientific-manageralist' and 'new-catastrophist' views tend to assume that the issue is one of global proportions, with complex scientific and political dimensions. In this context, to consider climate change as an *urban* problem may seem counter-intuitive. As a matter of GHG emissions, atmospheric conditions, global disruption, extreme events, international negotiations, rapidly increasing populations and national economies, climate change may at first glance appear to have little to do with the city. However, as the rest of this book sets out in detail, it is increasingly recognized that cities are not merely a backdrop against which these global processes unfold, but are central to the ways in which the vulnerabilities and risks of climate change are produced, and to the possibilities and challenges of responding to these issues. Understanding the urban nature of climate change requires that we examine, not only the ways in which the issue is being produced, reconfigured and contested within cities, but also how climate change has come to feature in the imagination and creation of urban futures.

The rest of this introductory chapter is divided into three sections. The first considers in more detail how and why we might view the global challenge of climate change in relation to the city. In the second section, the ways in which our visions of the implications of climate change and of urban futures shape our response to these issues are explored. As is clear from the discussion above, thinking about climate change evokes some image of, and relation to, the future. More or less implicitly, cities loom large in the scientific, political and cultural imagination of a climate-changed world. Thinking through the diversity of urban climate-change futures that emerge in this manner is instructive in terms of understanding how the city is regarded as both the cause of, and solution to, the climate crisis. Moving beyond these utopian and dystopian caricatures, however, requires a considered and detailed examination of just how and why climate change comes to matter in the city. The third part of the chapter outlines the rest of the book and the ways in which it seeks to address this challenge.

Before turning to these matters, a few issues of terminology are worth considering. Throughout the book, the terms 'urban', 'city' and 'cities' are used interchangeably. Broadly speaking, 'city' is usually used to refer to a specific place or locale that has urban characteristics (e.g. of density of population, social/economic activities, forms of public culture), but, beyond the political jurisdiction of urban local authorities (here termed municipalities), its boundaries are difficult to define. It is also vitally important to recognize the significant differences in the ways in which cities are encountered and experienced by different social groups and individuals – such as women, children, the economically affluent and mobile, or those who are economically and socially disadvantaged. Using the term 'urban' or 'city' can too readily mask those differences, although in an introductory book of this nature the use of such shorthand terms is inevitable. Throughout the book, the intention is to explore and examine how the multiple dimensions of climate change are experienced differently within and between cities, and to consider the implications for issues of social and environmental justice.

What has climate change got to do with the city?

It was in the early 1980s that the issue of climate change began to feature prominently on scientific and political agendas. Although early scientific investigations had shown that particular gases, when present in the atmosphere, could have a warming effect on the atmosphere by trapping incoming solar radiation, it was not until the 1980s that the potential implications of this physical phenomenon began to be realized by the international political community (Paterson 1996). Following a series of scientific meetings and political summits, in 1992 the landmark international agreement to address climate change, the *United Nations Framework Convention on Climate Change* (UNFCCC) was agreed. At the heart of the UNFCCC is the intention to stabilize GHGs in the atmosphere at a level that would 'prevent dangerous anthropogenic (human induced) interference with the climate system', while recognizing the need to adapt to the potential effects of climate change (UNFCCC 1992). Throughout this process, climate change was regarded as a problem of common property – although individual nation-states might be contributing to the problem, through their GHG emissions, the effects would be felt by all countries, and particularly in the least developed economies, who had not contributed significantly to the problem.

Resolving climate change, therefore, required an approach that could foster collective action for reducing GHG emissions while recognizing the different responsibilities of economically developed and less economically developed countries. Seen in this way, attention focused on how to achieve an international agreement that would bind nation-states to 'common but differentiated' targets and timetables (UNFCCC 1992), and, in 1997 the Kyoto Protocol was established as a means towards this end. The Kyoto Protocol committed thirty-eight industrialized countries to reduce GHG emissions by an average of 5.2 per cent below 1990 levels during the period 2008–12, and established a set of flexible mechanisms through which individual national targets could be reached. These included the provisions for establishing carbon trading between economically developed countries and the Clean Development Mechanism (CDM), through which economically developed countries could undertake projects to reduce GHG emissions in less economically developed countries as part of reaching their national targets. Since Kyoto, negotiations have continued over just how these provisions should be implemented and have grown in complexity as the climate-change agenda has grown to include different areas of activity – notably, how we should adapt to the effects of climate change, and how we might reduce the impact of deforestation on the atmosphere. These negotiations have also reflected the changing position of different nation-states over the past two decades, including the reluctance of the US to participate in an international agreement and the recognition that major industrializing economies, such as Brazil, India, China and Mexico, should have a role to play in meeting international targets for reducing GHG emissions. In 2009, the international community, global media, non-governmental organizations and many businesses with interests in responding to climate change gathered in Copenhagen for one set of negotiations, believing that a breakthrough in the impasse might be possible. The tangible disappointment that followed the apparent failure of these talks is a sign of just how much has come to depend on addressing climate change, and the extent to which it is considered as a *global* problem requiring *global* solutions (Bulkeley and Newell 2010).

There are, however, other ways of viewing the climate problem and the ways in which it could be addressed. If we step back from the idea of the global commons of the atmosphere and consider instead how, why and where GHG emissions are produced and the risks of climate change may be felt, a different set of processes, actors and possibilities comes to mind. The GHG emissions that are contributing to the condition of climate change do not arise from some uniform and invisible source, but, rather, are the product of the ways in which energy is used in our homes and cars, is used to make the things we consume and the goods we use, as well as through our management of the land and forests. It is these processes, taking place in a highly uneven manner across different national contexts, that create both current atmospheric conditions and the common but differentiated responsibilities for acting on climate change. At the same time, the impacts of climate change are not spread evenly, but rather occur in particular locations – such as those prone to flooding or drought – and affect different groups of society in radically different ways, depending on their capacity to cope with, and adapt to, risks. These ways in which climate change is closely intertwined with political economic structures and our daily lives is one of the reasons why it has proven so difficult an issue to address internationally. However, it also demonstrates that we can think about climate change differently – not as a *global* issue in the sense that it occurs in the same way across the world, but as one that has very different histories and geographies, varying across time and space and in its implications for economies and societies.

It is in this view of climate change, as shaped by diverse processes that vary not only between different nation-states but also within and across national boundaries, that the city comes into view (Table 1.2). Although the definition, nature and character of 'the city' as any singular entity are highly variable and contested, it can be shown that the twentieth century witnessed a rapid process of urbanization, particularly in the industrialized countries of Europe, North America and Australasia, and more recently in Asia, so that, by 2010, over half of the world's population lived within some form of urban context that is broadly understood as a 'city'. By 2030, it is predicted that almost 5 billion of the world's population of over 8 billion will live in cities, and this process of urbanization is likely to be most rapid in the least developed countries. Although large megacities will continue to grow rapidly, most urbanization is predicted to take place in smaller urban centres (UN-Habitat 2011: 2). As sites of rapidly growing population, cities have come to be seen as part of the climate-change problem. On the one hand, cities are now seen as potential sites of climate vulnerability. Because of the history of economic development, cities can occupy locations that are regarded as particularly vulnerable to climate change. The 2006 Stern Review, carried out for the UK government, finds, for example, that:

> Many of the world's major cities (22 of the top 50) are at risk of flooding from coastal surges, including Tokyo, Shanghai, Hong Kong, Mumbai, Calcutta, Karachi, Buenos Aires, St Petersburg, New York, Miami and London.
>
> Stern 2006: 76

At the same time, the rapid pace of urbanization is serving to produce its own forms of vulnerability within the city, as 'people arriving in already overstressed urban centres are forced to live in dangerous areas ... many constructing their own homes in informal settlements on floodplains, in swamp areas and on unstable hillsides', in areas that lack necessary infrastructures and basic services (UN-Habitat 2011: 1).

On the other hand, cities are regarded as critical to the production of the GHG emissions that are creating the risks of climate change in the first place

Table 1.2 Cities and climate change: part of the problem, part of the solution?

Cities as part of the climate problem?	Cities as part of the climate solution?
In 2010, more than half of the world's population lived in cities	Municipal authorities have responsibility for many processes that shape urban vulnerability and affect GHG emissions at the local level
By 2030, almost 5 billion of the world's 8.3 billion people will live in cities	Municipalities have a democratic mandate from local populations to address issues that affect the city
Cities have historically developed in locations that may be vulnerable to change, including in coastal areas and on rivers	Municipalities have a history of addressing issues of sustainable climate development
Rapid urbanization is creating significant urban challenges that will be exacerbated by climate change	Municipalities can act as a 'laboratory' for testing innovative approaches
Cities represent concentrations of economic and social activities that produce GHG emissions	Municipal authorities can act in partnership with private and civil-society sectors
Cities and towns produce over 70% of global energy-related CO_2 emissions	Cities represent high concentrations of private-sector actors with growing commitment to act on climate change
By 2030, over 80% of the increase in global annual energy demand above 2006 levels will come from cities in non-OECD countries	Cities provide arenas within which civil society is mobilizing to address climate change

Source: Adapted from UN-Habitat 2011: 91

(Bulkeley and Betsill 2003; UN-Habitat 2011: 91). The concentration of energy use within urban areas, in industry, transportation, and domestic and commercial buildings, means that cities are central to the ways in which GHG emissions, and particularly carbon dioxide, the most common GHG, are produced. The International Energy Agency (IEA) (2009: 21) has found that cities and towns currently consume over two-thirds of annual energy demand and 'produce over 70 per cent of global energy-related carbon dioxide emissions'. By 2030, their predictions are that, with 60 per cent of the world's population, cities and towns will consume over three-quarters of the world's energy demand, and that, 'over 80 per cent of the projected increase in demand above 2006 levels will come from cities in non-OECD countries' (International Energy Agency 2009: 21); that is, the vast majority of future increases in energy demand will come from those least developed and developing countries outside the Organisation for Economic Co-operation and Development (OECD).

As a resulting of the growing urban population and energy demand, cities are therefore seen as both the victim and culprit of climate change. Against this rather bleak picture, however, other views of the city as part of the *solution* to climate change are also emerging (Table 1.2; see also Bulkeley and Betsill 2003; UN-Habitat 2011: 91). Realizing the connection between climate change and existing forms of urban vulnerability, the nature of urban development and the daily use of energy requires recognition of the role that urban authorities – termed municipalities or municipal authorities in this book – have in shaping these processes. Municipalities have significant, although highly varied, roles in relation to urban planning, building codes, the provision of transportation and the supply of energy, water and waste services that shape existing patterns of vulnerability and the production of GHG emissions. Given these powers, and their democratic mandate as the local level of government, municipalities can therefore be seen as in a position to address the challenges of mitigating and adapting to climate change. Cities have a history of engaging with issues of sustainable development and can be centres for innovation, with the capacity to test new technologies and approaches, offering a means through which novel responses to climate change could be generated. Increasingly, cities are also home to a number of private and civil-society actors who are seeking to address climate change. In some cases, such actors are developing independent responses to climate-change mitigation and adaptation, but elsewhere municipalities can provide the basis for establishing partnerships with this diverse range of actors to approach these issues. It was this view, of cities acting on their ability to intervene in critical policy areas, democratic mandate, experience with sustainability planning and innovation, as well as their partnerships with other relevant actors, that led the mayors of some of the world's largest cities to call attention at Copenhagen to the importance of urban responses to climate change:

> We, the mayors and governors of the world's leading cities . . . ask you to recognise that the future of our globe will be won or lost in the cities of the world.
> (Copenhagen Climate Communiqué, December 2009)

Urban worlds and climate-change futures

For the group of the world's urban political leaders gathered at the 2009 Climate Summit for Mayors, taking place alongside the international Copenhagen climate talks, cities represented a battleground for our collective global futures. At stake, it seems, are competing views of what the city is and what it might become. As climate change has become a matter for popular culture and has entered into different forms of political, social and artistic consciousness, so too has the matter of urban futures become tied to our visions of what the climate problem is and how we might respond.

One such set of discourses has focused on the potential for climate change to cause catastrophe and the collapse of existing forms of civilization. The city takes central stage in these visions. In the 2004 film *The Day After Tomorrow*, rapid and abrupt climate change wreaks global havoc, with climate-related disasters striking Los Angeles, New Dehli and Tokyo and, centrally for the film's plot, flooding New York. In *Postcards from the Future*, an exhibition held at the Museum of London from October 2010 to March 2011, artists Robert Graves and Didier Madoc-Jones illustrated multiple forms of climate future for London, showing the fragility of the city in relation to the impacts of sea-level rise, abrupt glaciation, drought and climate migration. Beyond these images of the city as the somewhat unwitting victim of a climate-changed world are other urban dystopias in which the culpability of cities for climate change is writ large. Here, discourses and images of the city echo the official statistics of ever-growing urban populations and energy demand articulated by the United Nations and IEA, but the city is transformed from an abstract set of calculations to be seen as a site of insatiable consumption, greed and desire. In the 2010 novel by Ian McEwan, *Solar*, the central character describes his ambivalence on viewing the city from the air:

> Whenever he came in over a big city he felt the same unease and fascination. The giant concrete wounds dressed with steel, these catheters of ceaseless traffic filing to and from the horizon – the remains of the natural world could only shrink before them. The pressure of numbers, the abundance of inventions, the blind forces of desires and needs looked unstoppable and were generating a heat, a modern kind of heat that had become, by clever shifts, his subject, his profession. The hot breath of civilisation. He felt it, everyone was feeling it, on the neck, in the face.
> (McEwan 2010: 109)

In their discussion of the opportunities for, and limits to, urban responses to renewable energy, the IEA tells a 'Tale of two cities' (International Energy Agency 2009: 27–37). One, 'Bleak House', in which the impacts of climate change combine with urban dysfunction, including energy shortages, urban congestion, limited supplies of potable water, to create a bleak view of urban futures, a dystopia of urban neglect and decay. A second, 'Great expectations', represents a vision of 'the transition to a new, decentralised, decarbonised energy world: "it was the best of times"' (International Energy Agency 2009: 31), in which a range of new technologies that offer low-carbon energy are integrated into a new urban Utopia of a clean, vibrant and connected set of communities. This vision of a high-tech, low-carbon Utopia is increasingly associated with cities: a vision of plenty, continued economic prosperity and access to affordable energy for all. This is a view in which the individual elements of the city need to be brought together in a planned, coherent manner in order to respond to climate change. For Arup, an urban-development company, the challenge of climate change requires a holistic, integrated view of the city: 'creating a resilient city requires technical understanding of many individual elements, as well as the policy and regulatory frameworks that integrate them. Arup is well placed to help with it all' (Arup 2011a). In this vision, urban systems are able to be marshalled in response to climate change, creating a systemic and integrated response to challenges of both adaptation and mitigation. Such technical, often managerial, Utopias are of course not new in the urban arena – from ideas of the Garden City to Le Corbusier's visions of modernist urban planning, the urban has been an arena in which such ideals have been configured and contested. They rely on a particular sense of the urban as a system that can be known and is amenable to intervention. In other words, viewing urban responses to climate change in this manner involves a preconception of the urban as a relatively organized system, in which interventions can be readily planned and implemented. Various organizations, from IBM to Greenpeace, energy companies to governments, have started to create platforms and interfaces through which these discourses can be put to work to test new forms of urban future (Box 1.2).

Other utopian visions of the urban response to climate change are somewhat different. They are based on the idea of the urban as a place in which responses are emergent, coming from the activities and self-organization of independent actors, and where there are radical possibilities for living different urban lives in reconfigured urban economies. Less well established in mainstream discourses, this is a vision that might be termed 'decentralised engagement' (Foresight 2008: 117), in which multiple actors seek to develop measures that provide low-carbon energy and transport, or address urban vulnerability and foster resilience 'from the bottom up'. For some, this is a story of 'disruptive forms of innovation – cheaper, easier-to-use alternatives to existing products or services often produced by non-traditional players that target previously ignored customers' (Willis *et al.* 2007: 4). In terms of supporting adaptation, this is often a form of response focused on communities and recognizing that, 'planning and intervention design should use people's

> BOX 1.2
>
> ### Virtual urban responses to climate change: LogiCity
>
> > LogiCity is a fun interactive computer game with a difference. Aimed at young people under 25, it's a game set in a 3D virtual city with five main activities where players are set the task of reducing the carbon footprint of an average resident. As players work their way through the game they will pick up information about Climate Change and some of the main ways in which everyone is currently contributing to the emissions of the main greenhouse gas (CO_2) that causes Climate Change.
> >
> > It places individuals in a virtual environment and lets them experiment and experience the potential effect that individual actions can have on the planet before it's too late. Users can travel to the future to see the direct results of their actions in real time by quickly fast forwarding to the year 2066.
> >
> > The game is part of Defra's Climate Challenge programme to increase public awareness of Climate Change across the country. The National Energy Foundation, Logicom and British Gas are also providing some support to the game's development.
> >
> > (LogiCity 2012)

own ability and practice of experimentation and innovation as an entry point', while also working to overcome existing institutional and political constraints (Levine *et al.* 2011: ix). For others, the vision is of a more radical disconnect from existing norms of growth, consumption and urbanization, where, for example, new forms of food and energy provision are located in cities and managed through different kinds of collective action. While the Transition Towns movement (as discussed further in Chapter 7) is mobilized around the notion of community, this is increasingly being taken up in towns and cities:

> Transition Initiatives, community by community, are actively and cooperatively creating happier, fairer and stronger communities, places that

work for the people living in them and are far better suited to dealing with the shocks that'll accompany our economic and energy challenges and a climate in chaos.

(Transition Town Network 2012)

Although these different utopian views of urban responses to climate change differ significantly in how they view the role of technology, economy, government and community, they share a sense of the city as a space of possibility, potentially in the face of more far-reaching systemic change and collapse, and of the ways in which responding to climate change is not only a matter of addressing a seemingly global environmental challenge but also of reconfiguring urban systems and daily experiences.

More or less implicitly then, the city has come to figure in our political, economic and cultural imagination of climate change (Figure 1.1). Discourses of climate dystopias either play out across an urban landscape or place the city as central to unfolding catastrophe. The city is regarded as either the victim of the dangers emergent in the complexity of climate change, or as the culprit of the organized irresponsibility through which dangerous climate change is being produced. Utopian visions of the social and technical response to climate change are also conducted through different imaginations of the future city. One set of stories reveals different kinds of climate-controlled

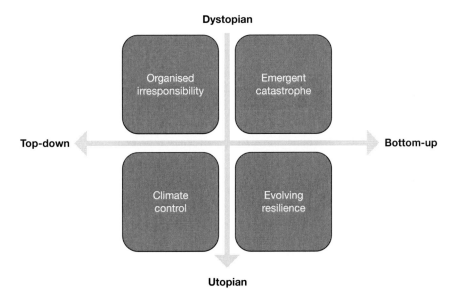

Figure 1.1 Imagining urban climate-change futures

future, achieved primarily through combinations of managed change and technical innovation. Another discourse shows the urban landscape as more emergent and fragmented, where the complexity of climate change leads, not to catastrophe, but rather to evolving forms of innovation and grass-roots alternatives with radical potential.

Chapter outline

Cities are now firmly on the climate-change agenda. The ways in which the effects of climate change are produced through and by cities and the manner in which our understanding of the urban is being challenged and reshaped by climate change are the subject of the rest of this book. Throughout the book, examples from different cities and communities are used to illustrate and explore the issues and arguments raised. Given the scope and range of material covered, readers are encouraged to use the references and suggestions for further reading to explore individual cases in more depth. This is also a fast-moving field, including many more valuable studies and examples than are referred to only in these pages.

In Chapter 2, we consider the issue of the *impacts* of climate change and the challenge of *vulnerability* in more detail. The chapter examines the challenges of creating accurate and useful information about how climate change might affect cities, and how and why we might come to understand urban vulnerability in relation to climate change. An in-depth case study of the issue of climate-change vulnerability in Cape Town illustrates these issues. Chapter 3 assesses how we can account for urban GHG emissions and the common but differentiated *responsibilities* for producing climate change that are emerging at the urban level. Some of the challenges of creating reliable and accurate models of urban GHG emissions relate to technical matters, but others concern just how we should draw the boundaries around cities and their wider connections to the processes of economic production and consumption. As the case study of Amman shows, these are important issues, particularly as cities seek to gain international financial support for taking action on climate change.

In Chapter 4, we turn to exploring how we might understand urban responses to climate change. It focuses on those urban responses that can be described as *governing*: that is, those that can be regarded as involving some degree of intervention in order to direct or guide the actions of others. The chapter outlines the multilevel context within which the history of urban climate-change governance has evolved, the different modes of governance that have been deployed, and the drivers and barriers that have shaped urban responses to date. The case of Melbourne is used to illustrate these processes in depth.

While the realities of many urban contexts mean that it is difficult to separate climate-change mitigation from adaptation, international convention, national policy and financial support for responses to climate change maintain this divide. At the level of the city, to date, significant differences can also be identified in the histories of mitigation and adaptation response, and in terms of the actors involved and initiatives that are being undertaken. Chapter 5 focuses on *climate mitigation and low-carbon cities*, first setting out the core terms and issues, before detailing how urban mitigation policy has developed, examining the evidence of how mitigation is taking place in practice, and considering the drivers and barriers shaping mitigation responses. Using the case of São Paulo, the ways in which this process of addressing mitigation takes place are examined in further depth. In Chapter 6, the issue of *adaptation* is addressed through a discussion of definitions and key debates in the field, analysis of the emergence of adaptation policy, and an examination of the practice of adaptation at the urban level and the drivers and barriers that have been critical in shaping this response. The case study of Philadelphia shows how adaptation policy and measures have only recently emerged on urban agendas and illustrates the challenging issues of vulnerability that they are seeking to overcome.

Whereas Chapters 5 and 6 place municipalities at centre stage, focusing on formal processes of policy-making and the sorts of intervention and measure that municipal authorities have undertaken, Chapter 7 starts from a different basis and examines the *experiments and alternatives* that are emerging as part of the urban response to climate change across a wide range of state and non-state actors. Despite significant challenges facing the development and implementation of urban climate-change policy, there is evidence of an increasing number of projects and initiatives taking place in cities that are apparently concerned with climate change. The chapter explores the reasons for this growth of 'climate-change experiments' and how we might understand their role in urban climate-change responses and considers four different types of experiment – as policy innovation, eco-city development, technological innovation, and the transformation of everyday practices. At the same time, different forms of climate-change response, concerned with issues of social justice, with resilience and with radical alternatives, can also be found taking place in cities. The chapter examines why and how such alternatives are emerging, and what their implications might be for how we think about climate change and the city. Chapter 8, in conclusion, summarizes the main topics and debates discussed in the book and considers their implications for issues of social and environmental justice, as well as the future of urban responses to climate change.

Discussion points

- Why and how can climate change be thought of as an urban problem?
- For one or more work of climate-change fiction – book, film, poem or exhibition, for example – analyse the ways in which the relationship between climate change and the city is articulated. Working in small groups or a class, compare and contrast the differences between different texts and discuss how these might be explained.
- Taking the example of a city with which you are familiar – your home town, somewhere you have visited, perhaps where you are currently living – create alternative scenarios for urban development in 2020, 2030 and 2050 under the four different forms of climate-change future set out in Figure 1.1. Which do you think is most likely, and why? In small groups or with your class, role play a town meeting in which a decision has to be made to follow one of these scenarios – which argument wins, and why?

Further reading and resources

There are now a number of classic books and important reports on the topic of cities and climate change that provide additional references to this text:

This is an edited collection that contains a number of chapters that address questions of vulnerability and adaptation in different developing countries:

Bicknell, J., Dodman, D. and Satterthwaite, D. *et al.* (2009) *Adapting Cities to Climate Change: Understanding and Addressing the Development Challenges*. Earthscan, London.

The first book to examine urban responses to climate change, this sets out the context for the emergence of urban approaches to mitigation and examines the processes of policy-making and implementation in six case-study cities:

Bulkeley, H. and Betsill, M. M. (2003) *Cities and Climate Change: Urban Sustainability and Global Environmental Governance*. Routledge, London.

This book examines how world cities are responding to climate change, providing an overview of the global situation and detailed examples from several major cities, including London, New York and Shanghai:

Hodson, M. and Marvin, S. (2010) *World Cities and Climate Change: Producing Urban Ecological Security*. Open University Press, Milton Keynes.

This is a comprehensive analysis of urban climate-change challenges, illustrated with lots of cases and examples:

Rosenzweig, C., Solecki, B., Hammer, S. and Mehrota, S. (2011) *Climate Change and Cities: First Assessment Report of the Urban Climate Change Research Network*. Cambridge University Press, Cambridge.

This is a comprehensive analysis of the relationship between cities and climate change, with a website that contains several background studies on individual cities, as well as links to other reports on urban sustainable development:

UN-Habitat (2011) *Global Report on Human Settlements: Cities and Climate Change*. UN-Habitat, Nairobi, Kenya. Online: www.unhabitat.org/content.asp?typeid=19&catid=555&cid=9272.

This is an important landmark report from the World Bank on the topic of cities and climate change:

World Bank (2010) *Cities and Climate Change: An Urgent Agenda*. World Bank, Washington DC. Further documents and analysis available online at: http://climatechange.worldbank.org/content/new-report-sees-cities-central-climate-action.

The following electronic resources provide different ways in which virtual cities are being created as a means through which to explore different ways of responding to climate change:

- www.greenpeace.org.uk/EfficienCity – virtual city used as an information portal on different energy technologies.
- www.electrocity.co.nz/Resources/ – New Zealand energy-company website designed for high-school students but useful for examining different kinds of alternative and the implications for decision-makers.
- www-01.ibm.com/software/solutions/soa/innov8/cityone/index.html – designed to simulate real-world decisions across different urban sectors and enable players to share their experience globally.

For one imagined view of the city and climate change, see *Postcards from the Future*: www.postcardsfromthefuture.co.uk/.

2 Climate risk and vulnerability in the city

> Climate change is not the only stress on human settlements, but rather it coalesces with other stresses, such as scarcity of water or governance structures that are inadequate even in the absence of climate change ... These types of stress do not take the same form in every city and community, nor are they equally severe everywhere. Many of the places where people live across the world are under pressure from some combination of continuing growth, pervasive inequity, jurisdictional fragmentation, fiscal strains and aging infrastructure.
>
> (Wilbanks et al. 2007: 373)

Introduction

As we saw in the Introduction, climate change is an increasingly important challenge for cities. The quote above, taken from the 2007 Fourth Assessment Report of the IPCC, suggests that this is because climate change will add stress to urban areas that are already under pressure from the effects of, for example, population growth, ill health, urban expansion, inadequate services, decaying infrastructure or persistent poverty. In urban areas, these risks are combined in distinct ways to create particular challenges, as well as opportunities, for living in the city. For others, climate change presents a more fundamental challenge to the organization and operation of socities. Understanding the impact that climate change will have in cities, therefore, means understanding how it will add to, or relieve, existing vulnerability. In its most fundamental terms, vulnerability relates to the exposure of people or places to risk. The extent to which cities may be vulnerable to climate change is, therefore, a product of the risks that they may experience – the impacts associated with different aspects of climate change – the ways in which they are exposed to those risks, and their ability to respond to risk (Box 2.1).

> **BOX 2.1**
>
> **Defining climate-change vulnerability**
>
> According to the IPCC,
>
>> *Vulnerability* is the degree to which a system is susceptible to, and unable to cope with, adverse effects of climate change including climate variability and extremes. Vulnerability is a function of the character, magnitude and rate of climate change and variation to which a system is exposed, its sensitivity and its adaptive capacity.
>>
>> (www.ipcc.ch/pdf/glossary/ar4-wg2.pdf: 883)

From the outset, it is clear that not every city will experience the risks of climate change in the same way, and that these experiences will vary significantly between social groups and individuals within any one urban context. As we saw in the Introduction, the predicted impacts of climate change range from sea-level rise to heatwaves, storms to droughts, and different places will be more or less exposed to these risks. Moreover, visions of climate change that suggest that it will lead to significant disruptions and conflicts also play out differently across urban contexts. At the same time, cities have highly differentiated abilities to cope with both the gradual effects of climate change and extreme events, and are more or less implicated in the potential conflicts and population flows that characterize more catastrophic views of climate-change futures. Whichever position is taken, importantly it is those cities in low- and middle-income countries that are likely to be the most affected by climate change:

> City-scale vulnerabilities are likely to be greater in developing country cities, primarily reflecting the fact that the population of these cities is often growing faster than their physical infrastructure capacity, and that their existing adaptation deficit to current climate variability as well as future exposure to climate change is greater than in developed countries.
>
> (Hunt and Watkiss 2011: 17)

In this context, this chapter examines how and why cities are vulnerable to the risks of climate change. The first part of the chapter outlines the impacts of climate change for cities, drawing distinctions between the direct and indirect effects that climate change may have, and considering the problem of assessing climate risks at the urban scale. The second part examines the ways in which vulnerability is created in cities by their location, by the process of urban development, and by social and economic conditions experienced in different urban communities. These issues are explored in detail in an in-depth case study of climate-change vulnerability in Cape Town.

The urban impacts of climate change

As highlighted in Chapter 1, climate change is predicted to lead to a range of effects that will vary from region to region, including sea-level rise, increased incidence of severe weather, changes in rainfall patterns leading to periods of flooding and drought, and increased temperatures and temperature extremes. This section first explores the nature of such risks for cities, before turning to consider how we can assess and evaluate the urban impacts of climate change.

Cities at risk?

At the urban scale, five types of climate impact are regarded as being particularly significant (Hunt and Watkiss 2011: 15):

- sea-level rise on coastal cities (including storm surges);
- extreme events (e.g. wind-storms, floods, heat extremes and droughts);
- health;
- energy use;
- water availability and quality.

Each of these risks has the potential to have both direct and indirect impacts for cities, some of which may be positive, for example in terms of creating new economic opportunities in tourist industries, or in the effect of rising temperatures on reducing the need for energy use in the winter and reducing winter deaths, but the majority of which can be regarded as negative (Table 2.1).

In some cities, climate change may serve to exacerbate existing risks, such as coastal flooding or heatwaves. In others, it may bring new challenges (Box 2.2). One particular example relates to those cities that are dependent for their water supplies or energy sources on glacial rivers. It is predicted that glaciers in many mountain ranges, including the Andes and the Himalayas, will have retreated significantly by the end of the twenty-first century, in turn leading to reduced river flows, the failure of hydroelectric-power schemes and a lack of available fresh water. Although this book focuses on urban responses to climate change, as this example shows, it is important to remember that the effects of climate change on cities will not only be felt within the urban

Table 2.1 Climate risks and impacts for cities

Climate risks	Examples of direct impacts in cities	Examples of indirect impacts for cities
Sea-level rise	Inundation and displacement of population Coastal flooding and storm surges Coastal erosion and loss of land Rising water tables and drainage problems Increased salinity of coastal environments Economic and leisure activities	Changing dynamics of ecosystems Changes to use of coastal zone Risks to marine economies
Extreme events	Damage to infrastructure systems, property, livelihoods and life from wind-storms, flood events, heatwaves and droughts	Risks to economic production chains Risks to urban food supplies
Health	Physiological effects of heatwaves and cold Changes in incidence of vector-borne diseases Physical- and mental-health impacts of extreme events	Risks to wider systems of health care and support
Energy use	Changes in winter and summer energy demand Increased use of air conditioning leading to brownouts	Risks to hydro-power energy systems Increased loss of transmission as temperature increases reduce energy supply
Water availability	Reduced precipitation and groundwater recharging limits water availability Retreat of glaciers reduces urban water supplies Increased demand for water as temperatures increase Reduction in water quality as river flow decreases	Risks to economic production chains Risks to urban food supplies

Source: Adapted from Hunt and Watkiss 2011

boundary, but also across the networks of water, food and energy supply that allow cities to function. Focusing on these networked connections, and their potential disruption, those who subscribe to the new catastrophism (Urry 2011: 36) suggest that the implications of climate change will go beyond discrete impacts on particular sectors and resources and are instead likely to cause widespread disruption, conflict and geopolitical upheaval.

The timing and extent of the predicted impacts of climate change are also important, but highly uncertain, factors in shaping the nature of their effects

---

BOX 2.2

Cities at risk

Mexico City

> With a predicted increase in mean temperatures of up to 4°C, together with a predicted decrease in mean precipitation of up to 20 per cent by 2080 . . . Mexico City can expect a more intense hydrological cycle, which is likely to affect the levels of the aquifers that still provide Mexico City with most of its fresh water. The expected increase in the evapotranspiration rate, along with decreases in the precipitation run-off and aquifer recharge rates will decrease the availability of fresh water for the city's inhabitants and economic activities.
>
> (Romero-Lankao 2010: 160)

Quito

> Quito is another city that will face water shortages as a result of glacier retreat. According to official studies, climate change is affecting mountain glaciers and water availability through a reduction in surface and underground water, the sedimentation of waterways, land use modification, the emergence of conflicts around water use, and increases in water use due to increased temperatures.
>
> (Hardoy and Pandiella 2009: 214)

> ### New York
>
> Accelerated sea level rise and exacerbated coastal flooding associated with climate change have been issues of critical concern for New York City and its surrounding region. With approximately 600 miles of coastline, this densely populated complex urban environment is already prone to losses from weather related natural catastrophes, being in the top ten in terms of population vulnerable to coastal flooding worldwide and second only to Miami in assets exposed to coastal flooding in the U.S.
>
> (Rosenzweig *et al.* 2011: 2)
>
> ### Toronto
>
> Climate changes are already being seen in Toronto. In the last decade, the City has been subjected to extreme heat, floods, drought, new insect pests, new vector-borne diseases and other problems made worse by climate change.
>
> (Toronto Environment Office 2008: 6)

in cities. On the one hand, sea levels and global temperatures are predicted to increase gradually over the course of the next century, while rainfall is expected either to decrease or increase, depending on the region involved. Climate models are able to provide a reasonably accurate regional picture of where such effects will be felt. The implications of these gradual, long-term impacts may therefore be relatively easy to assess. However, climate change is also predicted to bring a higher frequency and intensity of extreme events, such as storms and heatwaves, that have a much higher spatial resolution – i.e. that affect a small area at any one time – and are therefore difficult to capture with regional climate models. Understanding the nature and level of risk from such events is therefore very difficult at the urban scale. A further complicating factor is the potential that climate change may lead to 'tipping points' in the climate system that would create fundamental changes in current climatic regimes within a relatively short period of time. For example, scientists have expressed concerns about the potential for climate change to lead to disruption in the ocean circulation patterns in the North Atlantic, which could potentially lead to a relatively rapid (decade-scale) cooling of Europe and have wider implications for climate patterns globally. It is such potentially rapid and far-reaching impacts that give rise to some of the apocalyptic visions of future climate-changed cities discussed in Chapter 1. However, the current lack of certainty associated with predicting such impacts means that they are difficult to take into account in assessing urban climate risks.

Assessing the impacts

One of the greatest challenges cities face is in assessing the risks and their associated costs. Because of the difficulties in determining the scale and extent of climate impacts in cities, estimating the costs (and benefits) for cities that are likely to arise has also been challenging. Frequently, the approach adopted has been to consider the losses that occur from current events and to use the predicted trends of climate change to estimate future losses. However, this is challenging at the city scale. For example, although insurance companies have started to quantify the overall increases in damage that they expect from storms under conditions of climate change, limited information is available on individual cities (Hunt and Watkiss 2011). Estimates for New York and Miami, in the US, suggest that the maximum damages from one such storm or hurricane event would be in the region of 10–25 per cent of gross regional product (the economic product of the region) (Hunt and Watkiss 2011: 26). In terms of the costs of urban flooding, again some estimates are available. In a study of metropolitan Boston, US, researchers found that, if no adaptive actions were taken, the total losses would exceed US$57 billion by 2100, and that some US$26 billion could be attributed to the effects of climate change exacerbating this risk (Hunt and Watkiss 2011: 27). In 2005, in Mumbai, India, flooding also led to significant economic losses, estimated at some US$1.7 billion (Ranger *et al.* 2011: 141). Such estimates of current and future risks are useful in conveying the potential extent of the damages that climate change might cause, and in making the case for taking action to avoid these impacts. They are, however, fraught with uncertainties in terms of the challenges both of predicting future climatic change at the urban scale, discussed above, as well as of estimating in economic terms the loss of property, infrastructure, lives and livelihoods that accompany such disasters. Furthermore, focusing on the economic impacts of climate change may mean that those impacts that are difficult to quantify in monetary terms, including loss of life and impacts on health and housing conditions, are neglected. Given that there may be significant differences within and between cities in terms of whether the losses from climate change are felt in terms of assets or in terms of population, how the impacts of climate change are assessed becomes a matter of critical importance.

To overcome the challenges associated with using current events or global modelling to predict future climate impacts, a few cities have embarked on ambitious studies to use the predictions from global or regional climate-change models to make plausible predictions about the impacts of climate change at the local level. The New York City Regional Heat Island Initiative (NYCRHII) is one such case (Corburn 2009; Rosenzweig et al. 2011). Sponsored by the New York State Energy Research and Development Authority, the aim of the NYCRHII was to bring together scientists and policymakers to evaluate the potential impacts of climate change on the urban heat island (UHI) effect in New York and to assess the most cost-effective means of reducing vulnerability (Box 2.3). A regional climate model, the Penn State/NCAR Mesoscale Model, was used by the scientists in collaboration with the policy advisors to determine the likely incidence of future heatwaves in the city and to assess the impacts that different strategies – tree planting, creating 'green' roofs, and lightening surfaces such as pavements in the urban environment – would have on reducing their occurrence (Corburn 2009: 417). In order to do this, the model had to be downscaled: that is, it had to be made to work at a higher level of resolution than is normally applied to regional climate models. In this case, the model was downscaled 'by specifying variables at a modelling resolution of 1.3 km, rather than the 4 km or greater resolution at which [the model] is normally calculated' (Corburn 2009: 418), and also through the incorporation of local land-use data, such as vegetation cover, which would affect the nature of the UHI.

In his analysis, Jason Corburn (2009) found that this process was far from straightforward. Initial results from the model seemed to suggest that the proposed measures for reducing the effect of the UHI would have little impact on surface temperatures. This ran against the accumulated knowledge of the urban forestry department and other policy officials, who challenged these findings, suggesting that the modelling effort had not taken sufficient account of the microclimate of the city. With their advice, the model was downscaled further to a 10 m^2 grid in order to capture this fine detail. The resulting model suggested that surface lightening would be the most effective policy measure to implement (Corburn 2009: 420). This, however, was again controversial, as other policymakers with expertise in urban design and transport argued that the feasibility and costs of such measures had not been accurately assessed, again leading to changes in the modelling approach. The revised results suggested that urban tree planting would be the most effective strategy for addressing this particular climate impact. Yet again, this solution

> **BOX 2.3**
>
> **The urban heat island effect**
>
> The UHI effect works to elevate temperatures in urban areas with respect to their surrounding rural environments. It is caused by the fact that the surfaces commonly found in cities – buildings, roads, pavements and so on – absorb heat and reradiate it back into the city. The absence of green spaces or water within the city to provide shade and hold moisture exacerbates this effect. Although estimates vary, the UHI is thought to elevate temperatures in the city by 3–4° in comparison with rural areas. Because the UHI is produced through a combination of the extent of urbanization and the nature of the urban landscape and its morphology, as urbanization and densification increase in cities, it is thought that UHI will increase. In combination with the increased temperatures and heat events predicated to be associated with climate change, it is likely that cities will intensify more significant UHI effects in the future.

provoked controversy, as the 'Parks Department expressed concern that the modellers had overestimated the available area for planting street trees and thus the cooling potential of the intervention' (Corburn 2009: 422). Reflecting the ambiguities and uncertainties involved, the final report from the project notes that all three approaches to reducing the impacts of future heat stress in the city are comparable, but that tree planting may be done at the lowest cost (Corburn 2009, Rosenzweig *et al.* 2011).

At first glance, the case of the NYCRHII may appear to show the failure of science – the successive revisions of modelling efforts to integrate appropriate data and the continual process of negotiation with policymakers and their specific interests. In fact, it could be argued that it demonstrates the complex ways in which any assessments of climate impacts at the urban scale will need to be undertaken: assessments in which scientific and policy knowledge come together to produce relevant knowledge. This is because, in order to assess climate impacts at this level of detail, predictions of climate change that operate at global and regional scales need to be integrated with the knowledge of particular local climatic and urban circumstances. In this manner, such assessment processes may be able to provide municipal authorities with the levels of evidence and legitimacy that are required to act to address the problem (Chapter 6).

However, there are three important limitations to this approach to urban impact assessment and its widespread uptake. First, it is a process that relies on the availability of relevant local scientific knowledge, institutional capacity to integrate this knowledge into policy processes, and significant resources. This is reflected in the fact that, to date, those assessments of the impacts of climate change that have taken place have focused almost exclusively on sea-level rise, rather than considering other issues that may be more difficult to model and predict and therefore more resource intensive (Hunt and Watkiss 2011). A second, and related, issue concerns the extent to which such detailed levels of assessment are a necessary precondition for action to address the issue. As the case of urban heat in New York demonstrates above, the final conclusions of the report suggested that different interventions would all have an effect on reducing the risks of future heatwaves. Whether extensive climate modelling was required to reach this conclusion is debatable, although no doubt the process served to create political momentum and a shared agenda across the disparate agencies involved. Finally, such assessments are predominantly conducted through the bringing together of scientists and stakeholders. In such processes, those who may be affected the most by climate risks tend to have their views overlooked. An alternative approach has been undertaken by the UK charity Action Aid in cities in Africa. This participatory vulnerability analysis (PVA) includes policymakers and local community groups in assessing what makes them vulnerable, in this case to flooding, and how these issues can be addressed (Douglas *et al.* 2008). Not only does such a process provide a voice for those affected by risks, but it also reveals details about the causes of risk – in terms of exposure to hazard – that are not always accessible to experts or local decision-makers. By focusing on vulnerability, rather than the risk itself, such approaches may also provide a more appropriate basis for developing policy and adaptation options (see also Chapter 6).

As the first part of this chapter has shown, although in a global sense we are able to say with some certainty that climate change will have significant impacts on cities, pinpointing what these are and what they will mean is a significant issue. Furthermore, given, as we saw at the beginning of this chapter, that the significance of the impacts of climate change depends on the ways in which they interact with existing stresses and risks in cities, such analyses always have to be based on an understanding of the existing dynamics of vulnerability within cities – to which we now turn.

Climate change and urban vulnerability

The urban impacts of climate change are likely to be significant. Whether and how such impacts affect cities depend on the manner in which the infrastructures, buildings, people and activities taking place in cities are exposed and to what extent. This relationship between risks and exposure is defined as *vulnerability* (see Box 2.1). The vulnerability of cities to the impacts of climate change is often thought about purely in terms of their location – those cities that occupy coastal or riverside locations, for example, are regarded as being vulnerable to risks of flooding. However, the actual picture is more complex. Urban vulnerability is not only a result of geographic location, but is also shaped by the interaction between urban processes, daily lives and climatic risks. For some, vulnerability can be regarded as a result of the ways in which climate impacts exacerbate the risks caused by the underlying physical, economic and political processes that shape infrastructure systems, neighbourhoods and livelihoods in urban places. For others, vulnerability is also a profoundly social process. Different communities within the urban landscape are exposed to varying levels of risk and have diverse capacities for coping with climate impacts as a result of their unequal position within society. As we shall see below, these three views on the nature of urban climate vulnerability – in terms of location, in terms of place and in terms of community – create different views about who and what is at risk from climate change in cities.

Cities as vulnerable locations

Most studies of the risks of climate change to cities have focused on the impacts of sea-level rise and coastal flooding (Hunt and Watkiss 2011: 20). There are very good reasons why this is the case. First, coastal zones are highly populated areas, with large urban centres. By some calculations, 10 per cent of the world's population lives in the low-elevation coastal zone (McGranahan *et al.* 2007: 22), with '13 out of the 20 most populated cities in the world in 2005 being port cities' that are critical to the functioning of national and global economies (Hanson *et al.* 2011: 89). Second, the IPCC predicts that sea-level rise over the next 80 years is likely to be in the range of 0.2–0.6 m, suggesting that a degree of climate impact in the coastal zone is likely, but demonstrating that this is difficult to predict because of the complex feedback mechanisms between ice sheets, sea temperatures and changes to global atmospheric temperatures. Third, coasts, and particularly delta regions where cities are often located, are also subject to both natural and human-induced subsidence, contributing further to the risks of a rising sea level. In short, in the analysis of the impacts of sea-level rise on cities, it is their *location* in the coastal zone, coupled with growing levels of urban population, that is most often seen as placing them at risk to climate change.

In their study of the impacts of climate change on major port cities, Susan Hanson and her colleagues identified urban populations and assets – such as buildings, infrastructure systems, utilities, vehicles, material goods and so on – as exposed to this combination of rising sea levels, subsidence and population growth. Coastal cities are, of course, already exposed to risks associated with storm surges and flooding. Hanson's study shows that it is especially in delta regions that urban populations are exposed to these risks. With predicted changes to sea-level rise, coupled with subsidence and population growth, it is particularly the populations in rapidly growing cities in Asia that face increased exposure to climate change (Figure 2.1). At the same time, although it is in the developed countries that asset exposure to coastal and climatic risks is currently most significant, over the course of the next century, the increasing wealth of Asian cities means that they are also predicted to be at risk of significant loss of assets from the risks of climate change (Figure 2.2). In this account, trends in urbanization and economic growth mean that more people and assets are likely to be found in locations that are vulnerable to the combined impact of climate change and local subsidence. This shows us how the growth of urban populations is likely to lead to more climate vulnerability in the coastal zone. Such studies are more difficult to conduct for other climate risks, because of the uncertainty associated with, for example, predicting shifting patterns of rainfall, the intensity of storms or the occurrence of heatwaves. At the same time, although such approaches are very useful in setting out the broad trends that are likely to shape future urban vulnerabilities, they provide only a limited account of how urbanization produces climate risks and of how such risks are experienced by urban communities.

Cities as vulnerable places

A second perspective on urban vulnerability suggests that, rather than focusing on the location of cities, it is important to understand the ways in which the physical and economic development of cities contributes to their exposure to climate risk. In this perspective, climate impacts do not just happen *to* the city, but are fundamentally shaped *through* it – they are integral to the processes that create urban places. Several features of the urban landscape have been identified as exacerbating the exposure of urban places to climate

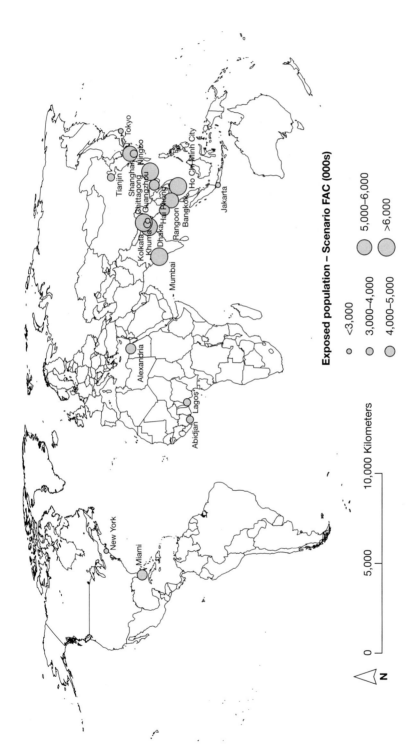

Figure 2.1 Map showing cities predicted to have populations most exposed to sea-level rise in the 2070s under conditions of climate change

Source: Hanson et al. (2011: 99)

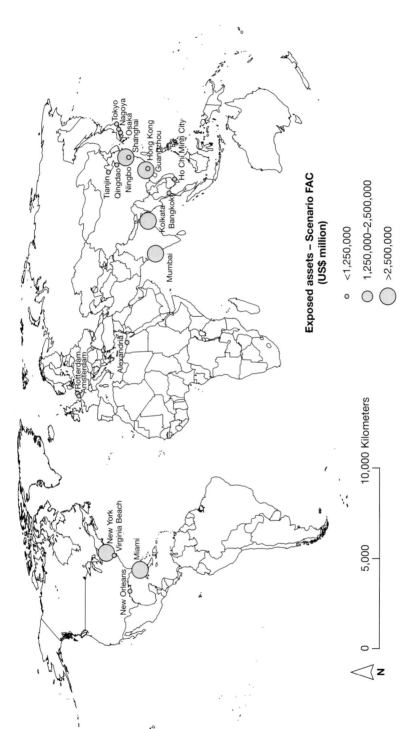

Figure 2.2 Map of the cities predicted to have assets most exposed to sea-level rise in the 2070s under conditions of climate change

Source: Hanson *et al.* (2011: 101)

risks. Perhaps most significant is the UHI effect, whereby the character of the urban landscape serves to keep city temperatures above those of nearby rural locations (Box 2.3). The character of urban water environments, with high levels of impermeable surfaces, which exacerbate the rapidity of stormwater flow through drainage systems, limited water-storage capacity and increasing demands for water use, means in turn that urban areas may be particularly prone to both flooding and drought. Urban areas also have infrastructure systems that may be particularly exposed to certain climate-related risks – for example, underground transportation systems and flooding – or are already experiencing significant strain, such as the sanitation systems and drainage networks in rapidly growing, informal settlements, or the electricity grid in established urban areas.

Importantly, the proximity and interdependence of infrastructure systems within urban areas mean that risks in one system can have negative impacts on another: for example, 'failures in wastewater treatment systems can increase requirements for water supply treatment, increase flood losses because of contaminated flood waters, and decrease availability of cooling water for power plants because of inadequate quality' (Kirshen *et al.* 2008: 106). The Climate's Long-term Impacts on Metro Boston (CLIMB) study, conducted between 1999 and 2004, found that climate change would lead to significant cross-system impacts. For example, in the energy sector, the study found that, by 2030, 'per capita energy demand could more than double compared to the 1960–2000 average, with climate change accounting for at least 20 per cent of the increase', and the remainder being due to demographic factors (Kirshen *et al.* 2008: 111). This increase in summer energy demand was predicted to have a number of other impacts in the urban arena, including rising levels of summer air pollution and associated ill health and impacts on water availability and on the quality of water (Table 2.2). In this analysis, it is the effects of climate change combined with a range of other economic and social processes operating across different urban infrastructure systems that produce urban vulnerability.

Understanding the processes through which urban vulnerability to climate-related risks is produced is therefore critical. Although most studies of future climate impacts have focused on the predicted levels of population and economic growth, it is important to recognize that vulnerability is also created historically and through the underlying political, economic and social dynamics of cities. Most significantly, in low- and middle-income cities, the expansion of urban areas has taken place through the development of areas previously thought to have been too vulnerable for human habitation. In Latin America, for example, urban expansion 'has taken place over floodplains or

Table 2.2 Climate impacts and the interactions between energy and other urban systems

	Energy	Health	Transport	River flooding	Sea-level rise	Water	Water quality
Energy	**Summer** More electricity demand. Also more brown outs and more local emissions **Winter** Less gas and heating oil demand	**Summer** Also decrease in air quality; higher morbidity and mortality **Winter** Also air-quality improvement	**Summer** Also if energy shortages; loss of rail service, loss of traffic signals, disruption of air traffic	NA	NA	**Summer** Also increased cooling water needs	**Summer** Also more cooling water will impact water quality (heat and blowdown)

Source: Adapted from Kirshen et al. (2008: 115)

up mountain slopes, or in other zones ill-suited to settlement such as areas prone to flooding or affected by seasonal storms, sea surges or other weather-related risks' (Hardoy and Pandiella 2009: 204). In these, often informal, settlements, a lack of land tenure, economic hardship, inadequate municipal investment and an absence of political rights combine to create conditions in which vulnerability to climate-related risks is an everyday occurrence. Such settlement patterns are also common in Asia and Africa. For example, in Lagos, urban development and the expansion of informal settlements in flood-prone areas means that, 'it is estimated that 70 per cent of the city's population lives in slums characterized by extremely poor environmental conditions, including regular flooding of homes that lasts several hours and that sweeps raw sewage and refuse inside' (Adelekan 2010: 433).

It is not only the extent and character of urbanization that are important here, but also the implications of urban development for the ways in which resources are used and managed in cities. In her analysis of the challenges of dealing with water in Mexico City, Patricia Romero-Lankao (2010: 157) argues that the city 'is already unable to cope with the types of climate hazards (e.g. floods, droughts) that global warming is expected to aggravate'. The supply, use and disposal of water and wastewater in Mexico City are the products of the history of urbanization and agricultural production in the area, with different administrations, from the Aztecs to the current municipal government, manipulating watercourses and drainage regimes while extracting ever more water. As a result, whereas the population of the city is expected to grow by some 17.5 per cent between 2005 and 2030, water availability is predicted to decrease by 11.2 per cent in the period 2007–30, even without the exacerbating effects of decreasing rainfall in the region (Romero-Lankao 2010: 166). At the same time, the city is prone to flood events, in part reflecting its valley location, but also as a result of the draining of natural sinks for water in the area, increased urbanization leading to faster run-off, and the inadequacy and disrepair of the sewerage and wastewater systems in the city.

The complex interactions between climate-change impacts, the historical development of the city and current urban dynamics mean that exposure to the risks of water shortage and of flooding is not spread evenly through the city. Reflecting the predominant historical pattern, the western, wealthier, areas of the city have more reliable access to piped water supplies and effective drainage systems, whereas areas in the east suffer from a lack of adequate infrastructure systems. Even though drainage systems have been built in the more recently populated, poorer parts of the city, they are inadequate, so that, for example, 'the poor functioning of the sewage system in Chalco and Netzalhualcoyotl ... results in chronic flooding with sewage in poor neighbourhoods' (Romero-Lankao 2010: 170). This is because 'decisions regarding infrastructure provision have been maintained and consolidated in a manner that benefits wealthy zones and contributes to a pattern of unequal spatial access to water' (Romero-Lankao 2010: 111). The landscape of climate risks in a city is therefore not merely a matter of which assets and populations happen to be located in vulnerable sites. Rather, climate risks are produced through the interaction between environmental and social processes, so that the risks that poorer areas face have been historically and systematically produced through urbanization. It is in this sense that the IPCC, in the quote that opens this chapter, suggests that climate risk 'coalesces with other stresses' that are produced by and through urban places. Understanding the causes, and consequences, of climate vulnerability means, therefore, that we need to look backwards, into the history of particular places, as well as forwards, in relation to future patterns of climate risk and processes of urbanization.

Cities as vulnerable communities

Acknowledging that climate vulnerability varies within and across urban places goes some way towards recognizing the complexity of the processes at work. However, it tells us little about how and why vulnerability is experienced by different members of urban communities. As set out above, it is the poorest places within cities that are most likely to be exposed to climate risks: the 'most vulnerable individuals, groups and places are those that experience the greatest exposure to hazards, but also those most sensitive or likely to suffer from exposure and with the weakest capacity to adapt' (Romero-Lankao 2010: 158). Urban poverty matters in terms of climate vulnerability, then, not just because of the inadequate shelter and infrastructure found in such areas of cities, particularly in low- and middle-income countries, but in as much as it also affects the degree of harm that such impacts might cause and the coping capacity of households and individuals (see the case study on Cape Town). However, although poverty is the most critical dimension of urban vulnerability, it is not the only one. Different communities within cities may be more or less vulnerable to the effects of climate risks because of other personal attributes, such as age and gender, the buildings within which they live and work, and the social networks of which they are part.

Although flooding poses significant risks to population and assets, as discussed above, the nature and distribution of this impact are more uneven, and the effects are broader, than such definitions reveal (Box 2.4).

CITY CASE STUDY

Climate change and vulnerability in Cape Town

Cape Town is one of South Africa's largest cities, with a fast-growing population of over 3.5 million people living on the Western Cape and with many of these residents living in informal settlements in conditions of poverty. The climatic conditions of Cape Town are considered to be similar to the Mediterranean, with dry, warm summers and wet and windy winters. Climate change is expected to have a number of impacts on these weather patterns that could exacerbate the problems associated with the climate of the Western Cape. These climate processes will increase the vulnerability of the city and particularly its poor communities. Furthermore, with over 300 km of coastline, a rise in sea level and increased storm surges will add to the complicated and interconnected risks and hazards posed by climate change. Vulnerability to climate change in Cape Town is wide ranging and provides a number of challenges for the municipal authority, communities and other urban actors around issues such as water supply, flooding, health, livelihoods and biodiversity.

The increased temperatures expected from climate change will create pressure on the urban water supply of the city, both in terms of less water being available and the demand for more water to cool households and businesses. Cape Town is expected to be the first city in South Africa to experience demand for water beyond its total capacity, owing to its fast growth and potential climate-change impacts. This could have a number of implications, including rising costs of water infrastructure and services, which will be felt most by poor households, increasing their vulnerability as costs become unaffordable. But increased vulnerability around water supply is not the only water-orientated challenge for the city in light of climate change. With the number of extreme weather events predicted to increase, storms will pose additional threats to Cape Town, putting communities at direct risk of flooding, as drainage systems are unable to deal with the storm-water, and tidal surges overcome sea defences. It is often the poorest, forced to live in hazardous areas such as the Cape Flats, which form part of the city flood plain, who are most vulnerable to these impacts. Previous flooding events have revealed these vulnerabilities, with families losing their homes, businesses and community facilities and being forced slowly to rebuild their lives, often with little external support available. Although the City of Cape Town spends time and resources working on disaster assistance for victims of flooding, in a city where there are many competing priorities and such a high proportion of poor people, these often prove inadequate and show the vulnerability of the city in dealing with natural disasters and extreme weather events.

Health vulnerabilities are also expected to increase as different climate-change processes impact on the city. For instance, the drier conditions will affect food security, with drought meaning that less food is produced and prices rise, affecting poor households and those with existing health problems such as HIV. The wet and cool winters pose serious health risks for Cape Town and its communities, with insufficient or inadequate housing unable to offer thermal protection from the damp and cold. Diseases and illnesses such as TB, pneumonia and colds are all expected to increase, putting additional strain on households and the city's capacity to support its communities. This vulnerability can affect a household in many ways, from the wage earner being unable to work, to children missing school, through to fatalities. The need to keep warm in the winter and the soaring cost of energy in Cape Town mean that many households suffer from fuel poverty, and, as a consequence, many residents use paraffin to heat their homes. However, this use of paraffin has serious fire implications, especially in the unplanned, informal settlements in which fire can spread dangerously quickly and hundreds of households are destroyed annually, as well as serious injuries and deaths. Fires from heatwaves are also predicated to increase as Cape Town experiences drier conditions during summer, with buildings and people threatened by hard-to-control wildfires. Again, those households that are already vulnerable through conditions of poverty are more likely to experience increased vulnerability from these processes.

The outlined vulnerabilities posed by climate-change processes threaten to impact across a range of different aspects of city life in Cape Town. One way to approach this increased vulnerability is to think about the impact on the livelihoods of residents in Cape Town. Livelihoods, the means by which people secure the necessities in life such as energy, water or economic opportunities, will be affected in many ways by this climate-related set of vulnerabilities. For many of Cape Town's residents, their livelihoods are often precarious and prone to disruption, meaning conditions of poverty are reinforced and making life more difficult for poor households. Academics, the municipality, community groups and others have been working on understanding the complicated series of vulnerabilities that impact on residents' livelihoods and offer a range of different policy and practical solutions to try and overcome these challenges.

The City of Cape Town has, over the last few years, developed policy responses to these issues. It has produced a number of strategies and reports that seek to understand these vulnerabilities and has begun to develop responses that seek to improve the resilience of communities to climate-change processes. The work includes assessments that have explored current and future vulnerabilities across various sectors, the development of strategies and plans to guide future actions, and the consideration of adaptation options that seek to increase resilience. What is clear is that the city faces a range of risks and hazards from climate change that threaten to pose a number of important questions about how vulnerable communities are to these processes. These include how these vulnerabilities interrelate with other social, economic and ecological stresses, and the priorities in a city with many challenges and only limited resources to tackle these vulnerabilities.

> Jonathan Silver, Department of Geography, Durham University, UK

Figure 2.3 Storm clouds over Cape Town
Source: Jonathan Silver

In communities across cities as diverse as Lagos, Dhaka and New Orleans, vulnerability to flooding was a result of the intersection of poorly developed or maintained infrastructure, long-standing patterns of urban development, social and economic circumstances, as well as access to the knowledge and resources that enable people to cope with risks. In other words, vulnerability is not only a result of the processes that produce particular urban places, but also varies between and within urban communities.

> **BOX 2.4**
>
> **The social impacts of floods in urban communities**
>
> Urban flooding is often associated with the loss of lives and assets, as well as with risks to health from waterborne disease. However, recent studies of the impacts of flooding on urban communities show that flooding has a broad range of other social and economic impacts that vary with wealth, age, gender, ethnicity and livelihood.
>
> ***Lagos***
>
> Alongside the direct impacts of flooding on urban roads, water supplies and the health of the population, residents in four of the poor communities in the city affected by flooding reported effects on their property and possessions, on the availability of drinking water, access to economic opportunities, as well as the stress of living with the ever-present risk of flooding, which led to mental health problems that were considered to be both far-reaching and difficult to address (Adelekan 2010).
>
> ***Dhaka***
>
> Flood risk is a common concern in this low-lying city, causing significant damage to urban infrastructure systems and to the built environment, particularly in the poorer parts of the city. As well as these direct impacts, which cause loss of life and ill health, a survey of residents conducted during the severe flooding of 1998 found 'at least 7.2 per cent of people had changed their occupation while 27.4 per cent were unemployed as a result of the flood', further compounding issues of poverty (Alam and Golam Rabbani 2007: 93).

For poorer urban communities, there are a range of factors that shape both exposure to risk and their capacity to cope with and adapt to such risks (see Chapter 6). Sheridan Bartlett (2008: 502) draws attention in particular to the ways in which children in such communities face increased climate risks, suggesting that their 'more rapid metabolisms, immature organs and nervous systems, developing cognition, limited experience and behavioural characteristics' contribute to a higher level of vulnerability. Drawing on evidence from current climate-related risks, she shows how children are more likely to lose their lives in extreme events than adults and more likely to suffer malnutrition and disease, which can leave them vulnerable to additional risks. In addition to their direct effects on children's health, climate-related disasters, such as flooding or droughts, can lead to separation, bereavement, displacement and disruptions to home and school, which in turn can affect mental and physical health, as well as opportunities later in life (Bartlett 2008; Table 2.3). Emergency shelter is often characterized by 'overcrowding, chaotic conditions, a lack of privacy and the collapse of regular routines [that] can contribute to anger, frustration and violence' that can be directed towards children (Bartlett 2008: 510). Alongside the risks of disasters, coping with low-level but persistent climate-related risks can also have effects on children, in terms of their health and levels of educational achievement. However, as Bartlett (2009) argues, this should not imply that children are only passive victims, for there is also significant evidence of their resourcefulness and ability to cope in the face of disaster. Instead, she argues that their views and knowledge should be taken into account by those seeking to adapt to the risks that climate change poses (Chapter 6).

Bartlett's study shows how vulnerability is created through the combination of poverty, age and the social and family networks of which children are a part. Like children, older people are also more likely to be vulnerable to some of the risks that climate poses. In Europe, the 2003 heatwave is thought to have led to a total of 15,000 excess deaths (Hulme 2009: 205), a disproportionate number of which occurred among older people. With predictions that such events are likely to increase in their frequency, intensity and duration in the future, understanding how and why heat vulnerability is created is increasingly important, particularly in cities where the UHI effect (Box 2.3) will exacerbate changing climate conditions (Wolf *et al.* 2010). The degree of vulnerability to heat risk is usually seen to be a product of both health factors, such as being older than 75 or suffering from chronic illnesses, and social relations, such as living alone or being socially isolated (Wolf *et al.* 2010). In this context, it is thought that access to what is sometimes called social capital – including engagement in social networks, trust, reciprocal relations of support – can reduce vulnerability. However, in a study of heat-related risk among elderly people in two UK cities, Johanna Wolf and her colleagues (2010) found that this was not necessarily the case. Instead, the key factors shaping vulnerability were the perceptions of risk among the elderly themselves and the strong beliefs among the social networks of which they

Table 2.3 Climate risks and urban children

Climate impacts	Risks for urban areas	Risks for health and household coping	Risks for children
Warm spells and heatwaves	UHI effect; concentrations of vulnerable people; increased levels of air pollution	Increased risk of mortality and morbidity; increased level of vector-borne disease; impact on those doing strenuous labour; respiratory problems; food shortages and food spoiling	Greatest vulnerability to heat stress for young children; high vulnerability to vector-borne disease and respiratory illness; highest risk of malnutrition with long-term consequences
Heavy rainfall events/ intense tropical cyclones	Increased risk of floods and landslides; disruption to livelihoods and businesses; damage to homes, possessions, infrastructure; often widespread displacement and disruption of social networks	Death and injury; increased risk of vector-borne and waterborne disease; decreased mobility and implications for livelihoods; dislocation; mental-health risks, especially from displacement and temporary shelter	Higher risk of death and injury; higher vulnerability to disease; risk of acute malnutrition; reduced options for play and social interaction; likelihood of being removed from school/put to work as household income drops; higher risk of neglect, abuse and maltreatment as a result of household stress
Drought	Water shortages, distress migration to urban areas, hydroelectric constraints, lower rural demand for goods/services, higher food prices	Increased food and water shortages, increased risk of malnutrition and food-/water-borne disease; increased risk of mental-health problems	Young children at highest risk from inadequate water supplies and malnutrition; risk of being removed from school/put to work as a result of a loss of household income, exploitation
Sea-level rise	Loss of property and enterprises; damage to tourism; damage to buildings from rising water table	Coastal flooding increases risk of death and injury; loss of livelihoods; health problems from increasingly salinated water	Highest risks of death and injury for children, highest health risks from salinization, long-term impacts of disease

Source: Adapted from Bartlett (2010: 504)

were part in the ability of the elderly to cope with heat. Among older people, there was little recognition of the risks that they faced, and indeed heat was regarded as something to be enjoyed by some of those involved in the study (Wolf *et al.* 2010: 47). At the same time, they found among the primary social contacts of the elderly –those people who offered support and care – that 'the primary narrative describes capable, competent and independent elderly people who are well equipped to take care of themselves in the face of heat risk' (Wolf *et al.* 2010: 49). Respecting the independence and capability of the elderly is clearly important; however, as this study shows, this can be misplaced where the understanding of risk is not adequate and can perhaps further contribute to vulnerability. In this sense, social capital may not always be a positive factor in reducing risk.

Conclusions

Climate change is predicted to have a range of impacts in cities, from sea-level rise to the increased incidence of storms, heatwaves, droughts and flooding. However, predicting the specific impacts that will occur in particular cities has proven to be scientifically complex and politically challenging, as different scientific disciplines and policy communities struggle to determine and work with uncertainty and to assess and quantify risk. The result is that, although we can say with confidence, at the global level, that climate change will have an impact on cities, determining just what those risks are, when they will occur and what impacts they will have is much more uncertain.

Assessing the potential impacts and implications of climate change also requires an understanding of the nature and dynamics of urban vulnerability. Although there is some consensus as to what vulnerability entails – in terms of the capacity of individuals, communities or systems to respond to risk – different approaches have been developed in order to understand and address climate vulnerability in cities. For some, vulnerability is primarily a matter of exposure to risk, and the key issues that need to be considered are the proximity of assets and populations to sources of risk. Others argue that understanding vulnerability means acknowledging the ways in which climate impacts are mediated through existing urban networks and processes and interact with other forms of risk. What matters by this account is understanding how the making of urban places may lead to vulnerability to the risks of climate change. Taking this argument further, others have argued that it is not the physical make-up of the city that matters in terms of producing vulnerability, but the social, economic and political processes that serve to marginalize and exclude some while privileging others. In this sense, urban vulnerability is produced through successive phases of urban development and is deeply entrenched within urban societies. Responding to climate-change vulnerability in cities, therefore, requires that persistent urban inequalities are also addressed.

Discussion points

- What kinds of knowledge are necessary for the assessment of the impacts of climate change in cities? What challenges does this pose for municipal authorities?
- Consider how two different climate risks – for example, heatwaves, flooding, sea-level rise, water shortage – might affect different vulnerable groups in one city. What are the similarities and differences in terms of how different groups are affected? What are the implications for how we understand 'climate vulnerability'?
- If climate vulnerability is shaped by other forms of economic and social vulnerability, is it useful to distinguish between them? What are the advantages of this approach, and what are the disadvantages?

Further reading and resources

The journal *Environment and Urbanization* contains up-to-date research on the issues and challenges discussed in this chapter.

The following contains a number of chapters that address questions of vulnerability:

Bicknell, J., Dodman, D. and Satterthwaite, D. (2009) *Adapting Cities to Climate Change*. Earthscan, London.

In 2011, *Climatic Change* published a special issue (volume 104) on 'Understanding climate change impacts, vulnerability and adaptation at city scale', which addresses some of the core debates raised in this chapter.

The Union of Concerned Scientists has a project setting out predicted climate impacts in California and the north-east US that shows in detail how climate change may affect these two regions; see www.climatechoices.org/index.html

Several cities have prepared their own assessments of the impacts of climate change, including:

- Cape Town: www.erc.uct.ac.za/Research/publications/06Mukheibir-Ziervoge%20-%20Adaptation%20to%20CC%20in%20Cape%20Town.pdf
- London: www.london.gov.uk/lccp/ and
- New York: www.nyc.gov/html/planyc2030/html/theplan/climate-change.shtml.

44 • Climate risk and vulnerability in the city

Information about climate vulnerability in particular cities is also available through the UN-Habitat's Cities and Climate Change Initiative:

- www.fukuoka.unhabitat.org/programmes/ccci/index_en.html.

Google Earth provides short tours of different places and ecosystems at risk from the impacts of climate change:

- www.google.com/intl/en_uk/earth/explore/showcase/cop15.html.

3 Accounting for urban GHG emissions

Introduction

> *Cities consume* over two thirds of the world's energy and account for more than 70 per cent of global CO_2 emissions, the most prevalent of the greenhouse gases.
> (C40 Cities Climate Leadership Group, May 2011a; emphasis in the original)

> Data drawn from the most recent report of the Inter-Governmental Panel on Climate Change (IPCC) suggest that cities are not responsible for 75–80 per cent of greenhouse gas emissions. Carbon dioxide from fossil fuel use accounted for only 57 per cent of global anthropogenic greenhouse gas emissions in 2004, and a very large proportion of non-carbon-based greenhouse gas emissions are not generated within cities.
> (Satterthwaite 2008a: 539–40)

As we saw in the Introduction, owing to their growing populations and rising levels of energy and resource consumption, cities are increasingly seen as a significant contributor to the climate-change problem. In 2006, the Stern Review, commissioned by then UK Chancellor of the Exchequer Gordon Brown, suggested that 'by some estimates, urban areas account for 78 per cent of carbon emissions from human activities' (Stern *et al.* 2006: 457). In its 2008 World Energy Outlook, the IEA calculated that urban areas account for more than 71 per cent of energy-related global greenhouse gases currently and suggested that this could be expected to rise to 76 per cent by 2030 (International Energy Agency 2008). These figures, and others like them, have given rise to an increasingly widespread understanding that cities have significant responsibilities for addressing the causes of climate change, as articulated above in the quote from the C40 Cities Climate Leadership Group, one of the leading organizations working in this arena. However, despite the seemingly straightforward nature of such claims and their well-intentioned aims of supporting urban action on climate change, others have challenged their validity and the assumptions that lie behind them. As David Satterthwaite (2008a) explains, too often levels of energy consumption, GHG emissions and carbon dioxide emissions are confused, creating considerable uncertainties about just what role cities may play in contributing to the climate-change problem and how they might respond.

Concerns have also been raised that such broad-brush statements about the responsibilities of 'cities' – as if they were all the same – tend to mask the significant differences between them in terms of energy consumption and GHG emissions production. Although current indicators suggest that GHG emissions have been concentrated in cities in the developed countries, which are members of the OECD, the IEA finds that, 'by 2030, over 80 per cent of the projected increase in demand above 2006 levels will come from cities in non-OECD countries' (International Energy Agency 2009: 21). What is more, there are large variations in the contribution of cities within each of these broad groups to global GHG emissions, both between and within individual countries. In addition, the actual definition of what counts as an urban area is far from clear, serving further to complicate the matter of working out just what role cities play in contributing to the climate-change problem.

Despite these challenges, accounting for urban GHG emissions remains an important task. In the international arena, climate politics is guided by the general principle of 'common but differentiated responsibilities' – the recognition that, although all countries have a common interest in avoiding dangerous climate change, their duty to address the problem is differentiated according to their contribution to the problem (UNFCCC 1992). At the urban level, therefore, understanding the sources and potential future growth of urban GHG emissions is important in terms of considering the level and nature of different urban responsibilities for acting on climate change. Such knowledge may also help in the development of appropriate solutions. Understanding the contributions being made to GHG emissions by different types of activity in different cities can help us understand how and why they vary and how they might be reduced. For example, understanding the differences between transport systems in terms of their contribution to GHG emissions in different urban contexts may help us to identify the most appropriate alternatives to reduce GHG emissions in a range of urban conditions. Furthermore, knowledge about the GHG emissions being produced in particular urban places and through different processes within the city is a critical step towards gaining access to the increasing amount of finance that is being directed towards addressing climate change. Accounting for and evaluating the urban contribution to GHG emissions is, therefore, a critical part of understanding responses to climate change in the city.

The rest of this chapter is divided into two main sections. The first section considers the challenges of measuring and monitoring GHG emissions at the urban level and introduces the differences between taking a *production*-based or a *consumption*-based approach to accounting for GHG emissions. With these challenges in mind, the second section compares the differences emerging between different cities in terms of their contribution to GHG emissions and outlines some of the factors that explain the very different profiles of cities emerging worldwide. It also considers the question of the consequences of such assessments, and in particular what they imply for where responsibility for taking action lies. A case study on Amman explores in detail the connections between accounting for GHG emissions and accessing new sources of carbon finance, and the challenges and problems with such approaches. The conclusion reflects on the main issues raised by the chapter and identifies some critical questions to consider and suggests further reading.

Assessing the urban contribution to climate change

As the urban contribution to climate change has come into focus, an array of approaches, methodologies and tools have been developed for assessing GHG emissions at the local level. In parallel with developments at the international and national level for assessing GHG emissions, these tools have become more sophisticated and robust over the past two decades. These approaches primarily adopt an approach that allocates GHG emissions to the place in which they were *produced*. That is, they seek to account for those GHG emissions that are produced as a result of processes that occur in particular urban areas. Of these approaches, 'the most widely accepted methodology for measuring emissions within local government boundaries has been developed by Local Governments for Sustainability (ICLEI)' as part of their Cities for Climate Protection (CCP) programme (UN-Habitat 2011: 35; see also Chapter 4). Following the development of various different methodologies for measuring emissions across different parts of the CCP network, in 2009 the *International Local Government GHG Emissions Analysis Protocol* was produced, following a set of principles developed for use in measuring commercial GHG emissions (Box 3.1). As with previous tools developed by ICLEI, it allows municipalities to calculate their own GHG emissions, as well as to assess those for their local area (termed 'communities' by ICLEI), the result of which is an *inventory* of current GHG emissions.

> **BOX 3.1**
>
> **Principles for the design of the International Local Government GHG Emissions Analysis Protocol**
>
> > **Relevance:** The greenhouse gas inventory shall appropriately reflect the greenhouse gas emissions of the local government or the community within the local government area and should be organized to reflect the areas over which local governments exert control and hold responsibility in order to serve the decision-making needs of users.
> >
> > **Completeness:** All greenhouse gas emission sources and activities within the chosen inventory boundary shall be accounted for. Any specific exclusion should be disclosed.
> >
> > **Consistency:** Consistent methodologies to allow for meaningful comparisons of emissions over time shall be used. Any changes to the data, inventory boundary, methods, or any relevant factors in the time series, shall be disclosed.
> >
> > **Transparency:** All relevant issues shall be addressed in a factual and coherent manner to provide a clear audit trail, should auditing be required. Any relevant assumptions shall be disclosed and include appropriate references to the accounting calculation methodologies and data sources used, which may include this Protocol and any relevant Supplements.
> >
> > **Accuracy:** The quantification of greenhouse gas emissions should not be systematically over or under the actual emissions. Accuracy should be sufficient to enable users to make decisions with reasonable assurance as to the integrity of the reported information.
> >
> > (ICLEI 2009: 7–8)

For an individual municipality, calculating either its own or community-level GHG emissions requires establishing the *scope* of emissions to be included, the sectors from which these will be calculated and the level of data that will be used to complete this task. In keeping with international and national conventions, municipalities have focused on measuring and monitoring Scope I

> **BOX 3.2**
> **Defining Scope I, II and III GHG emissions**
>
> Following the Greenhouse Gas Protocol, a voluntary carbon accounting tool developed by the World Business Council for Sustainable Development and World Resources Institute, GHG emissions from organizations are categorized in three groups or 'scopes':
>
> **Scope I – Direct emissions:** direct emissions resulting from activities within the organization's control. Includes on-site fuel combustion, manufacturing and process emissions, refrigerant losses and company vehicles.
>
> **Scope II – Indirect emissions:** electricity and heat. Indirect emissions from electricity, heat or steam purchased and used by the organization.
>
> **Scope III – Indirect emissions:** other. Any other indirect emissions from sources not directly controlled by the organization. Examples include: employee business travel, outsourced transportation, waste disposal, water usage and employee commuting.
>
> Source: The Carbon Trust 2012

and Scope II GHG emissions, rather than also including Scope III (Box 3.2). It is this focus on emissions that arise within a jurisdiction or as a direct result of the energy consumed within that area that has led to a form of accounting that directs attention to where GHG emissions are produced. In terms of the sectors that are included in such inventories, analysis suggests that these can vary significantly between cities. For example, an analysis conducted by the Carbon Disclosure Project (CDP) of the municipal emissions inventories of twenty-eight of the largest cities in the world found that, although 93 per cent included GHG emissions from buildings in their analysis, only 39 per cent included information on water supply, and just 14 per cent contained data on employee commuting (Carbon Disclosure Project 2011: 17). There is also significant variation in the sort of data used to analyse urban GHG emissions. In setting out the International Local Government GHG Emissions Analysis Protocol (Box 3.1), ICLEI suggests that there are three 'tiers' of data that can be used for this purpose, each of which contains a trade-off between accuracy, availability and comparability:

> Tier 1 is the basic method, frequently utilizing IPCC recommended country-level defaults, while tiers 2 and 3 are each more demanding in terms of complexity and data requirements. Although tiers 2 and 3 are considered to be more accurate, there is a trade-off between the effort involved in obtaining the information and the benefit of having it. Local governments analyzing greenhouse gas emissions from their municipalities should use the highest practicable tier.
>
> (ICLEI 2009: 23)

A more detailed and accurate inventory of urban GHG emissions is therefore dependent on the sorts of data included. Whereas most municipalities that have sought to calculate their emissions have relied on the use of protocols and the input of 'Tier 1' data, as with the ICLEI model above, others have developed specific approaches for their urban areas. This alternative approach involves the gathering of local data (e.g. on energy supply and use, transport patterns and building stock) and the 'bottom-up' development of an emissions inventory. One such example is the ClimateCam model developed in Newcastle, Australia, where recent and real-time data on community-wide GHG emissions are provided at a district level and made publicly available on the Internet, on a billboard in the city and in a weekly news report (Box 3.3 and Figure 3.1).

BOX 3.3

ClimateCam, Newcastle, Australia

ClimateCam is the world's first greenhouse gas speedometer. It provides an accurate measuring tool to track the greenhouse gas emissions for the City of Newcastle, NSW, Australia. ClimateCam will provide monthly consumption data and greenhouse gas emissions from the following sectors:

- Electricity
- Water
- Waste
- Tree planting
- Transport
- Gas
- Beach water quality
- Total emissions.

ClimateCam will provide monthly estimates for consumption data and greenhouse gas emissions from gas and transport in Newcastle, based on the data available.

(Newcastle City Council 2012)

Figure 3.1 ClimateCam billboard, Newcastle
Source: Photo provided by City of Newcastle, Australia

Despite the growing popularity of GHG emissions inventories and evidence that municipalities have been able to use them to create a viable basis for municipal action, several key challenges remain (Allman *et al.* 2004; Lebel *et al.* 2007; Sugiyama and Takeuchi 2008; UN-Habitat 2011). One critical issue is the availability of data. In many cities, data concerning the nature of the energy standards of buildings, daily travel patterns, energy consumption and so on are simply not routinely collected. In cities where a large proportion of the population lives in informal or illegal settlements, the lack of data is a particular challenge. Even though such areas are likely to contribute little to the overall GHG emissions of a city, underestimating the population and economic productivity of such areas may serve to inflate the production of GHG emission per capita or per unit of GDP in such cities, painting a false picture of their contribution to the problem. For example, in a study of the implications of urbanization for the carbon footprint of Chiang Mai, Lebel *et al.* (2007: 111) found that 'the consequence of these various processes on overall carbon stocks, fluxes and balances could not be estimated with much precision in this study, because of limitations of adequately disaggregated or relevant local data on emission factors'. Even where data are collected, a large quantity is not available in the public realm as it is held by private utility companies who regard it as commercially sensitive. This has been a critical issue for local authorities in the UK, who have long campaigned for access to locally relevant data on energy supply and consumption (Allman *et al.* 2004). Furthermore, it may be difficult to ascertain just what kinds of data are needed at which spatial scales. As ICLEI acknowledges, information about the flow of energy and materials is mostly easily obtainable at the national level, so that, as the 'spatial area of analysis is reduced to city or municipality, the accuracy of an analysis may be further reduced due to the difficulty of tracking the movement of materials and energy across jurisdictional boundaries' (ICLEI 2009: 9).

The issue of measuring and monitoring the flow of energy and materials across jurisdictional boundaries points to another core challenge facing efforts to develop urban analyses of GHG emissions – how and where to draw boundaries. Defining urban boundaries, particularly with regard to informal settlements and migrant populations, remains contested. In the case of GHG emissions, drawing tightly defined urban boundaries may be necessary for the purposes of calculating inventories, but has the potential to be misleading. For example, one area where different conventions are adopted relates to how transportation is accounted for within GHG emissions inventories. In most cases, urban-community GHG inventories include transportation within a particular geographical boundary, but may not take account of those journeys that begin outside this area or that originate in the city but involve travel beyond its boundaries. This is particularly the case for air transportation, which is rarely included in inventories, but also relates to suburban and interurban commuting by car or by rail. In effect, the challenge here is one of deciding where the production of GHG emissions associated with an activity that, by its very nature, crosses boundaries should be allocated.

A more fundamental challenge comes, however, when the very basis of GHG inventories on the *production* of GHG emissions is considered. The production focus of inventories means that they do not routinely account for the GHG emissions that are embodied in the goods and services that we consume which are not produced in the urban places where we live. Most significantly, the GHG emissions embodied in the materials (e.g. for construction), products (e.g. clothing, consumer electronics) and food that we consume, and in the waste that we dispose of outside the urban boundary (e.g. recycling of aluminium), are not allocated to the city in which these are consumed. Rather, a production-based accounting system allocates these

GHG emissions – and, implicitly, the responsibility that goes with them – to those cities that produce such goods and services (Dodman 2009; Satterthwaite 2008a). Although this reflects conventional forms of financial accounting, which attribute the production of economic activity to the places in which it is generated, given that production chains are now globalized and that the GHG-intensive parts of these processes are frequently located in urban centres in the Global South, the effect is to attribute the responsibility for these embodied emissions primarily to developing countries (UN-Habitat 2011: 58).

The complexities of the methodology adopted, scope, sector and level of data included, combined with the challenges of data availability, boundary definition and the exclusion of 'consumption'-based GHG emissions have created a patchwork of information about the urban contribution to GHG emissions. In a recent assessment of the contribution of C40 cites, the Carbon Disclosure Project found that:

> Variations in size, makeup and methodology combine to create massive differences in the amount of emissions reported by each C40 city, with the spread between cities up to ten-fold. 4 cities report total community emissions of 5 million metric tons CO_2-e, whereas 2 other cities show community emissions to be greater than 50 million metric tonnes CO_2-e.
> (Carbon Disclosure Project 2011: 20)

Such are the complexities that a recent report from UN-Habitat finds that:

> It is impossible to make definitive statements about the scale of urban emissions. There is no globally accepted standard for assessing the scope of urban GHG emissions – and even if there was, the vast majority of the world's urban centres have not conducted an inventory of this type.
> (UN-Habitat 2011: 45)

Despite these serious reservations, efforts to measure, monitor and verify urban GHG emissions continue apace. In May 2011, ICLEI and C40 announced that they were collaborating to develop a new global standard for 'accounting and reporting community-scale greenhouse gas emissions', arguing that a 'common approach will help local governments to accelerate their emission reduction activities whilst meeting the needs of climate financing, national monitoring and reporting requirements' (ICLEI 2011). As this statement reveals, the increasing interest in cities as a means through which to address climate change is attracting the interest of global carbon finance and other levels of government that will depend on urban GHG emissions being accounted for in such a way that it fits into internationally accepted protocols (see the case study on Amman below). Although such an integrated approach holds much promise, it is unclear whether simply creating a common measurement tool will address the underlying challenges surrounding data collection, determining urban boundaries, and choices over allocating emissions to production or consumption activities. As discussed further below, overcoming these issues may require a different way of thinking about urban responsibilities for the climate-change problem.

CITY CASE STUDY

Calculating carbon finance for Amman

Since the Kyoto protocol was first adopted, carbon financing has become a key aspect of finding market-based solutions for decreasing anthropogenic GHG emissions. The premise of the CDM is to award emissions-reduction credits to projects in developing countries that can be exchanged and used by industrialized countries to meet their targets under the Kyoto Protocol. An international panel awards the credits for projects, which are then financed by international institutions. For example, since the Kyoto ratification in 2005, the World Bank has established multiple carbon finance 'products', including the Umbrella Carbon Facility (UCF) and the Prototype Carbon Fund (PCF); funds resulting from bilateral agreements to purchase emissions reductions with countries such as the Netherlands, Denmark, Spain or Italy; and strategic funds such as the Forest Carbon Partnership Facility (FCPF), the Community Development Carbon Fund (CDCF) and the BioCarbon Fund (CFU 2010a). These funds have financed both national initiatives for carbon reduction and large infrastructure projects led by private actors. However, numerous criticisms have been raised regarding the types of activity that can be accessed through carbon finance and the potential of these interventions to move beyond business-as-usual scenarios.

Cities have not accessed carbon finance on a wide scale. This is manifest in the relatively low number of CDM projects in urban areas. International agreements and policies have been generally slow in recognizing the potential of local, rather than national, governments to take climate-change action. Subnational governments often need the mediation of national governments to access international finance and to participate in international governance debates, and climate change is no exception. However, the main difficulties faced by cities in accessing these funds – in comparison with private actors – are related to practical aspects of carbon finance (CFU 2010b). First, local governments, whether alone or in partnership with other actors, may be unable to deal with the high administrative and transaction costs of setting up CDM projects. In general, high transaction costs have meant that carbon finance has been confined to large-scale projects, which local authorities with limited resources may not be able to undertake. Second, local authorities may find it difficult to measure the direct local benefits of projects for GHG-emissions reduction (the benefits of which are better observed at the global level), and this will make it difficult to justify these expenditures at the city level. Cities have promoted the idea that emissions reductions could be achieved by ensuring that mitigation projects are associated with co-benefits, providing additional services to the city (for example, in public-transport projects). When this occurs, however, the challenge lies in demonstrating the 'additionality' of the measures being taken to the CDM board, that is, to demonstrate that the project adds substantially to what would already be taking place in the city in the absence of carbon finance.

International organizations have proposed institutional reforms to make carbon finance accessible at the city level. The World Bank Institute has established the Carbon Finance Capacity Building (CFCB) programme to provide advice and support on carbon finance in mega-cities in the South. In addition to providing targeted advice, the World Bank Carbon Finance Unit has proposed a citywide approach to carbon finance. In this approach, the city is taken as a unit for GHG-emissions reductions. Emissions-reduction activities in different sectors (energy, transport, solid waste, water, wastewater and urban forestry) are then added together to develop a single city programme for GHG-emissions reductions that can be used to access carbon funds (both through international finance and voluntary markets). This approach relies on the idea of developing a 'programme of activities' to enable the aggregation of emissions reductions from different activities. A citywide approach is currently being piloted in Amman, where the Greater Amman Municipality (GAM) has developed a Green Growth Program (Amman City Wide Clean Development Mechanism Program) that adopts a programmatic approach to carbon finance. The Program has three interrelated objectives:

1. to improve the urban environment while contributing to the climate agenda;
2. to improve the cost efficiency of municipal services;
3. to mobilize additional sources of revenues through carbon markets.

This programme of activities in Amman will aggregate carbon emissions from activities in the waste, energy, transport and forestry sectors (Table 1). The Green Growth Programme will be combined with the Metropolitan Growth Plan which aims at radically changing the urban form of the city with initiatives to renovate the downtown (Wadi Amman) and implement 'mixed-use street' principles, with resonates with the notion of building compact cities (see Chapter 5).

Representatives at the WB argue that the Program has been made possible by the presence of individuals within GAM offering both technical and political leadership, recognising both the potential challenges of a doubling population by 2020 (up to 5 million) and the increasing scarcity of water and energy resources. Previous experience working with the WB in a municipal waste program helped GAM to access the right contacts at the WB, in order to develop the Programme of Activities for the city. The GAM's plans also contemplate multi-level governance arrangements between the Jordan government and the GAM, for example, through the Jordan Renewable Energy Fund. They argue that the Amman Green Growth Program is 'the first of its kind in the world' and it is expected to open the door to other cities to adopt a similar approach to carbon finance.

Figure 3.2 Amman City
Source: Photo by Siegfried Atteneder

The main strength of the Programme of Activities is its flexibility. Eligibility is established at the beginning of the program so that new projects can be added without independent approval. This reduces the proportion of transaction and administrative costs for each independent intervention. There are technical challenges in transforming citywide interventions into a quantifiable regime for GHGs emissions reductions. The focus on performance contracts requires projects to be measurable, which restricts the capacity of cities and other actors to obtain CDM credits for innovative projects or projects where carbon measurement techniques have not yet been developed. For example, the plans within Amman's Green Growth Program are linked to carbon finance, but they also contain objectives and activities (see Table 3.1) that may not contribute directly to offsetting carbon emissions but, nevertheless, improve the livability and sustainability of the city. The GAM representatives have stated their commitment to finding alternative sources of revenue where carbon credits are not applicable. Nonetheless, the problem of quantifying emissions reductions and the multifunctional 'co-benefits' accrued through sustainable-development actions may serve to limit the possibility of adopting a city-based programmatic approach to carbon finance that can move beyond a narrow range of sectors and activities.

Table 3.1 Outline for a programme of activities for Amman

Program component	Action	Notes
Urban transport	Bus rapid transit system	Emphasis on limiting motorized transport that will match the priorities of GAM's planning agenda, focusing on 'mixed-use streets'
	Light rail transit system Fuel switch for buses and other public vehicles	
Municipal waste	Land filling with recovery of landfill gas	Projects will continue previous collaborations with the WB to improve the efficiency of the waste-management system in the city
	Recycling of plastics Slaughterhouse waste-to-energy	Managed through a public–private partnership, it attempts to demonstrate state-of-the-art technologies

continued . . .

Table 3.1 ... continued

Program component	Action	Notes
Urban forestry	Urban agriculture	Motivated by food-security concerns. Productive land concentrates inside the spatial boundaries of the city. The main limitations relate to Amman's ongoing problems with water scarcity
	Plantations in urban and peri-urban areas	They could use recycled water, having additional benefits for municipal waste
Sustainable energy	Energy-efficient street lighting	Focus only on replacing bulbs. Although it is regarded as 'easy' and 'simple', at the moment it is still in a pilot stage
	Promotion of residential CFL usage	The National Research Centre leads a pilot project in 700 households
	Solar water-heating systems for households	Focus on the involvement of the private sector, because the main limitation is the capital investment
	Wind farm	The GAM has selected suitable locations, but the project is still to be developed

Source: Vanesa Castan Broto

Vanesa Castan Broto, Development Planning Unit, University College London

Urban difference and the drivers of GHG emissions

Despite all of the challenges of measuring and allocating GHG emissions to specific urban areas, efforts continue to try to determine the different contributions that different cities are making to this global problem. Recognizing these differences can be an important means of understanding the relative significance of different types of city in terms of overall global GHG emissions and can also help us to understand the underlying drivers that lead to GHG emissions. This, in turn, can be an important step towards identifying the sorts of action that cities might take to reduce, or mitigate, their GHG emissions. This section first assesses the differences that can be found between urban areas in terms of their GHG emissions, before considering what this might tell us about the urban processes and dynamics that are responsible for these emissions. We then turn to reflect on what these differences and drivers mean for how responsibility for taking action should be considered at the urban level, and the implications for creating urban responses to climate change that are both sustainable and fair.

Urban diversity and different contributions to GHG emissions

As will now be clear, statements about the particular contribution of a city to GHG emissions need to be understood in the context of the limits of data availability, differences in measurement approaches and so on. Nonetheless, such measurements do provide a useful way of starting to think about how and why cities might be responsible for global GHG emissions. Such allocations are routinely done at the international level, where the clear differences between the contributions of different countries to global GHG emissions levels are now reasonably well understood. One useful way of viewing this information has been created in the form of a map showing the proportion of 'global-warming potential', an index that is used to aggregate the combined effect of different GHG emissions (Worldmapper 2012). Presented in this way, it is clear that it is the wealthiest countries and those that are developing their economies most rapidly that currently contribute the most significant proportion of GHG emissions to the global total. It is, therefore, not surprising to learn that

> 18 per cent of the world's population living in developed countries account for 47 per cent of global CO_2 emissions, while the 82 per cent of the world's population living in developing countries account for the remaining 53 per cent.
> (UN-Habitat 2011: 45)

Bringing an urban perspective into this analysis begins to add some detail to the general picture. Analysis conducted by the World Bank suggests that

> the world's 50 largest cities, with more than 500 million people, generate about 2.6 billion tCO_2e annually, more than all countries, except the United States and China. The top 10 greenhouse gas emitting cities alone, for example, have emissions roughly equal to all of Japan.
> (World Bank 2010: 16)

Although, as set out in Chapter 1, the possible contribution of cities to energy-related carbon dioxide emissions may be as high as 70 per cent, this analysis suggests that there are particular types of urban setting – in this case, large metropolitan areas that are often centres for economic and political activity – that may be particularly critical in terms of their contribution.

However, determining which cities are the most significant in terms of their contribution to GHG emissions depends both on how such an attribution is made and on the basis by which it is measured. Using the data from the World Bank's analysis, it is possible to draw up at least three different lists of the 'top ten' cities in terms of GHG emissions, even when just using a production-based perspective (Table 3.2). In Table 3.2, the first column lists the top ten cities in terms of their overall contribution to GHG emissions. This shows, mirroring the national picture, that it is the largest cities of the US and China that are contributing the most to global GHG emissions. The second column, however, shows a slightly different picture. Based on per capita GHG emissions, that is, emissions per person, it indicates that other cities, which may not be as large – such as Moscow, Toronto and Dortmund – are contributing significantly, if we consider the amount that each person living in these cities is responsible for producing. The third column provides some indication of a top ten list of cities in terms of the amount of GHG emissions that are produced per unit of economic production – measured here in terms of GDP. This provides an indication of the energy and carbon intensity of officially measured economic activities in different cities. Interestingly, this measurement provides a very different list of cities, including those in some developing countries whose total contributions to GHG emissions globally are dwarfed by wealthier cities, such as Lagos and Kinshasa. Given the problems of data availability and accounting discussed above, which are particularly acute in such cities, these figures do, of course, need to be treated

Table 3.2 Which are the top ten? Three different ways of evaluating the global contribution of cities

Total GHG ($MtCO_2e$)		Total GHG per capita ($MtCO_2e$/cap)		GHG per GDP ($kyCO_2e$/US$bn)	
New York	196	Moscow	15.4	Tianjin	2,316
Tokyo	174	St Petersburg	15.4	Beijing	1,107
Moscow	167	Los Angeles	13.0	Shanghai	1,063
Los Angeles	159	Chicago	12.0	St Petersburg	971
Shanghai	148	Miami	11.9	Moscow	922
Osaka	122	Philadelphia	11.9	Lagos	893
					(total = 27 mT)
Beijing	110	Shanghai	11.7	Bangkok	799
Chicago	106	Toronto	11.6	Riyadh	726
Tianjin	104	Dortmund	11.6	Tehran	560
London	73	Tianjin	11.1	Wuhan	554
Bangkok	71	Bangkok	10.7	Kinshasa	598
					(total = 6 Mt)
Miami	65	Beijing	10.1	Istanbul	384

Source: World Bank 2010

with caution. It is also worth noting that, in focusing on the formal economy, for which GDP contribution can be calculated, other informal economies, which provide livelihoods and services for the urban population and which may be much less carbon intensive, go unrecorded, exaggerating the GHG emissions intensity of the total economy. Nonetheless, it provides us with an important indication that the carbon intensity of economic production is higher in those urban economies that supply many of the goods that we currently consume.

Given the diversity of urban conditions globally and the fact that the term urban can be applied to all sorts of scale of human organization – from small cities and towns to the sorts of mega-city that appear in Table 3.2 – many analysts find that the per capita emissions measure can provide the best indication of the contribution that particular urban conditions and lifestyles are making to GHG emissions. Such a measure, they suggest, can more easily show the differences between and within cities than aggregated measures of total emissions, which are primarily determined by the size of populations and the nature of energy supplies in a given city. Within the United States, for example, there are significant differences between New York, which is the largest total contributor to GHG emissions, and other cities in terms of per capita emissions. New York's per capita emissions are '40 per cent lower than Houston's per capita emissions' (World Bank 2010: 16) and 'are half those for Denver, [at] 10.5 tCO_2e versus 21.5 tCO_2e', which is 'mainly attributable to New York's greater density and much lower reliance on the automobile for commuting' (Hoornweg *et al.* 2011: 4).

Furthermore, using per capita GHG emissions as an indicator of the contribution of particular urban places can highlight the differences between urban and rural conditions, and how the contributions to overall GHG emissions vary within one city. Comparing the per capita GHG emissions of urban residents and national averages, research suggests that, in developed economies, urban conditions usually work to create lower per person GHG emissions. Particularly significant differences have been found between suburban areas and inner cities (Box 3.4). A 2009 study in the United States found that, 'an average household in 48 major metropolitan areas generates up to 35 per cent less greenhouse gas emissions when located in the city than when located in the corresponding suburb' (World Bank 2010: 17). However, looking at GHG emissions in emerging and developing economies, the opposite is often the case, with GHG emissions higher than average in many large cities in these countries. In China, for example,

> Shanghai's emissions are 12.6 tCO_2e per capita, while national emissions are 3.4 tCO_2e per capita. This reflects the high reliance on fossil fuels for

electricity production, a significant industrial base within many cities and a relatively poor and large rural population, and hence a lower average per capita value for national emissions.

(Hoornweg et al. 2011: 3)

There are also significant differences between cities in China. Analysis by Shobhakar Dhakal (2010: 76) finds two different groups of cities, one of 'energy-intensive cities . . . largely located in the central and western parts of China, which house energy-intensive industries and lie in climatically cooler areas' and another of 'less energy intensive cities . . . that are located in the eastern part of the country . . . closer to the coast and have strong service industries and a relatively warmer climate'. An analysis undertaken in South Africa similarly shows that the relationship between GHG emissions and urban conditions varies according to the type of city and urban economy involved, so that 'non-industrial towns and cities have average per capita emissions of 3.4 tonnes; "metros" 6.5 tonnes; and industrial towns and cities, 26.3 tonnes' (UN-Habitat 2011: 49).

Despite some broad trends, therefore, there are important differences in the contributions that cities are making to global GHG emissions. Globally, we can say that it is large cities that are contributing the biggest total amounts of GHG emissions, but this masks wide variations in the nature of this contribution and its underlying drivers. Given the global picture, it is also not surprising to conclude that, on the whole, richer cities use more energy

---

BOX 3.4

What are Toronto's per capita emissions?

On average, residents in the city core produced 6.42 tCO_2e per capita compared to 7.74 tCO_2e per capita for residents in the surrounding suburbs. However, there were pockets within the city core that produced emissions as high as those in the suburbs; these census tracts represented wealthy neighbourhoods, characterized by high automobile use and older, inefficient homes. The lowest emissions were 1.31 tCO_2e per capita for a dense inner-city neighbourhood with good access to public transportation. The highest emissions were 13.02 tCO_2e per capita in a 'sprawling' distant suburb.

(Hoornweg et al. 2011: 8)

and produce more GHG emissions than poor cities (World Bank 2010: 17). At the same time, however, concentrations of energy-intensive economic production in some cities mean that some cities in developing countries appear to have significant per capita emissions. Often, the goods that are produced through these carbon-intensive practices are not destined for consumption within that city, but rather are exported globally for other consumers. Perhaps the most important conclusion from this analysis is that there is 'no one single factor that can explain variations in per-capita emissions across cities, instead the variations are due to a variety of physical, economic and social factors specific to the unique urban life of each city' (World Bank 2010: 24). Although accounting for GHG emissions can be one step towards uncovering these factors, understanding these dynamics requires further consideration of urban conditions and lifestyles.

Urban dynamics and the drivers of GHG emissions

At the most fundamental level, the amount of GHG emissions that are produced in any one urban area is related to the type of energy being used – whether it is 'carbon intensive' – and the total amount of energy consumed – be that in buildings, motor transport and/or the supply of water and sanitation services. Although there are different measures of the 'carbon intensity' of sources of energy available, broadly speaking, electricity that is produced from coal or diesel generators creates more GHG emissions per unit than that from gas, while nuclear power and renewable energy (such as hydro, wind and solar) are the least carbon intensive. Fuels derived from oil, petrol and diesel, are more carbon intensive than biofuels, such as ethanol, and heating that is based on oil is more carbon intensive than gas systems, but here, too, other renewable energy sources, including heat pumps and solar hot water, are the least carbon intensive. Differences in the type of energy used in various cities therefore have an impact on their overall contribution to GHG emissions. In Australia and Canada, where electricity is primarily generated through coal-fired power stations, GHG emissions per capita are higher than in European countries, where electricity is often provided by gas, nuclear and renewable energy. Differences between per capita GHG emissions in cities in Latin America, where there is a high proportion of hydro-powered electricity, and those in China, where coal is used to create electricity, account in part for the diversity of urban contributions to GHG emissions seen in Table 3.2.

Although the type of energy used to power cities and urban lifestyles can account for some of the diversity discussed above, it is not a sufficient explanation. Cities in the same country with the same sort of energy mix have distinct levels of GHG emissions. As we saw above, urban, suburban and rural conditions also affect GHG emissions per capita in some striking ways. In addition to the type of energy, it is the amount that is used that matters. This is true whether you take a *production*- or *consumption*-based view of the ways in which GHG emissions should be attributed to cities, although the emerging picture of the drivers of GHG emissions is very different in each case.

Drivers of production-based GHG emissions

In seeking to account for the differences between how and why cities produce GHG emissions, analysts have turned primarily to four further explanations. First is a set of issues that we can term the basic geographies of cities. These include the range of weather conditions experienced by cities and associated demands for heating and cooling, as well as processes of population growth (or decline), and levels of wealth and service provision. For example, distinctions have been drawn between 'colder-climate' Chinese cities and those in the more temperate regions as the basis of explaining their varied energy needs. As also discussed above, where urban populations are poor, total energy consumed tends to be lower, although this varies significantly within cities. Usually, such features of cities are taken for granted as shaping the natural demand for energy. However, diverse levels of the use of, for example, heating, between cities with similar climates, suggest that there is more to this picture than simple cause-and-effect relationships between certain attributes of cities and their GHG emissions. Second, as discussed above, the nature of urban economies is also important in shaping GHG emissions locally – cities that are dependent on energy-intensive industries are likely to contribute higher total and per capita GHG emissions than those whose economic base is in the service sector.

A third set of factors that are thought to be important in shaping energy use in cities comprises those of urban form and density. The layout of cities, distances between areas of work, home and leisure, together with the ways in which buildings are spaced and clustered, is now thought to be an important factor shaping urban contributions to GHG emissions. On the whole, it is argued that the denser a city, the lower its contribution to climate change will be. This is because of two key sets of relationships between urban form and energy use. In formal urban areas with adequate shelter, where urban development is denser, the need to heat buildings is reduced. Compact dwellings tend to share more walls, take up less floor space and create a stronger 'urban heat island' (Chapter 2), reducing the amount of energy required to provide warmth. Their layout also lends itself to the implementation of more efficient systems for providing energy, such as district heating and/or cooling networks. In urban areas where informal and inadequate shelter dominates, any such benefits are seldom realized, and instead it can be argued that high levels of density pose a hazard to health and increase climate vulnerabilities (see Chapter 2; UN-Habitat 2011: 54).

Beyond the individual buildings, research suggests that the pattern of urban development is important in shaping mobility. Where urban development is less dense, and particularly in areas where 'sprawl' is growing, there are limited options for using public transport or for walking and cycling. As density declines, comparative analysis of mobility patterns in different parts of the world suggests that car use increases. The correlation between the use of private motorized transport and reduced GHG emissions suggests that this is an important factor shaping energy use in cities (Newman and Kenworthy 1999). There is, however, no simple relationship at work. In an analysis of the changing demand for motorized transport in India, Jun Li found that:

> the dynamics between urban structure such as road infrastructure and transport energy consumption is considerably complicated. Investment in road infrastructure is largely a response to the growth in car ownership; the growing demand resulting from increased car ownership is driving road infrastructure further. Improved infrastructure is also pushing up the demand for cars, which, in turn, makes it attractive to purchase new cars.
> (Li 2011: 3506)

Rather than determining car ownership and use, this analysis suggests that cultures of mobility are also important in shaping urban form. Rather than viewing suburbs or urban sprawl as the causative factor shaping the growth in car ownership and GHG emissions, this suggests that we need to consider suburban growth as something that is *enabled* through, as well as contributing to, car-based mobility cultures (Paterson 2007). From this perspective, particular kinds of urban form are the product of underlying political and economic processes that foster certain types of development. Seeking to tackle urban GHG emissions through creating more compact forms of urban development, without addressing these root causes, may not be very successful (Chapter 5).

A fourth set of factors shaping urban energy use relates to the nature of the built environment itself. The age and type of the built environment, structures of ownership and the functions for which it is used all go to shape the ways in which GHG emissions are produced from the built-environment sector. For example, in the UK, more than 25 per cent of total CO_2 emissions are 'from homes. In 2005, 53 per cent of these domestic carbon emissions were from heating space, with another 20 per cent from water heating, 22 per cent from lights and appliances, and 5 per cent from cooking' (Foresight 2008: 58). In the UK, the rate at which buildings are replaced is low, meaning that, by 2050, '65–70 per cent of the dwelling stock in existence is likely to have been built before 2000' (Foresight 2008: 59). In London, the proportion of energy used in the domestic built environment is considerably higher than the national average, at 38 per cent, suggesting that how energy is generated and used in this sector is of particular significance for cities (Bioregional 2009: 11).

In seeking to determine the factors that shape the production of urban GHG emissions, the geographies, economies, density and built form of cities have been examined. Too often, these issues are regarded as natural attributes of cities from which causal explanations can be developed. Whether this is in relation to the number of cooling degree days, the structure of local economies, urban densities or the existing building stock, the contribution of GHG emissions is regarded as an outcome of these particular urban attributes. As a result, such explanations often fail to take account of the histories and cultures, politics and economics of urban development and the ways in which these dynamics shape contemporary energy production and use. In this way, an analysis that focuses too heavily on the production of GHG emissions within particular cities not only may attribute 'responsibility' to cities in a way that does not account for their role in global networks of production and consumption, but it may also fail to account for the processes that have structured urban development and that create particular path-dependencies. Turning to analyse the ways in which processes of consumption shape GHG emissions may provide one way of starting to assess these dynamics.

Drivers of consumption-based GHG emissions

At the most fundamental level, rather than being a result of the sheer number of people in any city, consumption-related GHG emissions are driven by income and, therefore, vary within and between cities, according to the distribution of income across different neighbourhoods. As outlined above, for the poorest people and the poorest cities, levels of GHG emissions are orders of magnitude lower than for wealthy individuals and urban areas, reflecting a lack of access to affordable and reliable energy services (UN-Habitat 2011: 53). At the same time, population dynamics – such as ageing societies, the growth of single households and smaller family units in individual houses – also have an effect on the total amount of GHG emissions produced in any city.

Population and wealth provide us with some baseline differences upon which to base understanding of how patterns of consumption might affect urban GHG emissions. However, the total contribution of a city to GHG emissions is driven by what has been termed the *urban metabolism*, that is, by the throughput of materials and energy in a city and the wastes that it generates (Kennedy *et al.* 2010). Such analyses can be used to provide a comprehensive 'carbon footprint' for cities, showing the different forms of resource use and consumption that contribute to GHG emissions. For example, an analysis of the consumption emissions for London shows that the contribution to GHG emissions from the city doubles when consumption – including food, consumer goods, personal services such as banking and entertainment – is taken into account (Bioregional 2009). Such an analysis suggests that a different set of dynamics is critical to urban contributions to GHG emissions, including accepted norms about clothing, diet and leisure.

Examining the food sector in more detail, the London study found that red meat contributed to 27 per cent of these GHG emissions, fruit and vegetables 15 per cent, and dairy products 12 per cent (Bioregional 2009). Reducing GHG emissions from these forms of consumption is complex. On the one hand, the GHG intensity of the products we buy and eat is shaped by a whole set of different actors operating across the supply chain. The same London study reports that, '43 per cent of the emissions come from the agriculture stage, 15 per cent from manufacturing, and 20 per cent from transport, storage and distribution. These three stages account for over three quarters of the emissions associated with food' (Bioregional 2009). As a result, there is little that individual consumers in London may be able to do to reduce the carbon footprint of particular goods through their own behaviour. Rather, reducing the GHG emissions contribution of urban food consumption may require action by other actors across the supply chain, new forms of food consumption, including more locally sourced food, and changes to diets (see Chapter 7). Making such changes is far from straightforward. Taking just one of these sectors as an example, a report by WWF finds that:

> Dairy products are pervasive, enjoying a high level of penetration, both in volume and frequency terms, and form a core part of eating occasions throughout the everyday life. This implies that dairy products are deeply engrained in everyday habits and cultural behaviours. Over 90 per cent of people consume milk and the majority do so daily. The three most popular uses of milk are on cereal, in tea or coffee, or on its own as a cold drink, especially by children. Cereals are eaten most days of the week by 85 per cent of the population and tea and coffee is drunk an average of three times a day by 70 per cent of the population. . . . Milk consumed as ice cream is eaten as a treat, a snack during the summer period or as a desert option, by over 70 per cent of people, on a weekly to monthly basis.
>
> (Jackson *et al.* 2009: 26–7)

Not only is dairy consumption ubiquitous, such cultural practices of consumption are routinized in everyday life. Figure 3.3 shows routine 'dairy moments' in the average UK day. Here, the role of dairy in everything from holiday and teatime treats such as ice cream to the typical British cup of tea is clear.

It is not only food and consumer goods that are shaped by culture, habit and what we might expect from a 'good life'. As Elizabeth Shove (2003) has so persuasively argued, our use of more invisible services – energy, water and waste – is also structured by societal norms surrounding comfort, cleanliness and convenience and the seemingly mundane, technical infrastructures through which they are delivered. In the urban context, GHG emissions are therefore structured not simply by urban geographies, economies, density and the built environment, but through the social and technical systems of provision that deliver energy, water and waste services, which are both shaped by, and serve to structure, urban conditions (van Vliet *et al.* 2005).

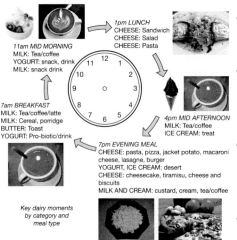

Dairy products are eaten in an array of formats throughout the day; on their own as a nutritious snack or as the main component in popular British meals such as porridge, pizza and sandwiches. Therefore major shifts in meals formats and occasions would be required to reduce dairy consumption, where substitute products are not available.

- According to Mintel (2006) and TNS (2007), the three most popular uses of milk are for tea/coffee, cereal and drinking it on its own as a cold drink. None of these uses commonly involve cooking, which reflects a broader decline in home cooking.

- The key meal time for cheese is lunchtime, though it is associated with cooking themes and special occasions. Over 70% of people eat cheese sandwiches. There are also clear links to use in Italian cooking. Christmas is an important event in the cheese calendar, placed on the menu by over 1/3 respondents in Mintel survey.

- The majority of people who consume yogurt do so once a week. Yogurt drinks tend to be eaten less often then yogurt may be used as snacks or as breakfast products to promote good health. Heavy users of yogurt use it every day for cooking and as a desert. Medium users tend to keep it in the fridge as a snacking option and desert option.

- Butter is increasingly being used as a medium for cooking, especially meats.

Sources: Cheese, Yogurt, Yellow Fats Market Intelligence, Mintel 2005, Impact of consumption, Defra report, 2007.

Figure 3.3 Daily dairy moments

Source: Jackson *et al.* (2009: 28). Reproduced with kind permission from WWF and Imperial College London
Image sources: yogurt by Oxytousc, available at http://commons.wikimedia.org/w/index.php?title= User:Oxytousc&action=edit&redlink=1 (Creative Commons); latte by Mortefot (Creative Commons); cheese salad by cyclonebill, available at http://www.flickr.com/people/23178876@N03 (Creative Commons); ice-cream by Lumen GmbH (Creative Commons); tea by Bubble36288, available at http://commons. wikimedia.org/w/index.php?title=User:Bubble36288&action=edit&redlink=1 (Creative Commons); macaroni cheese by Hoikka1, available at http://commons.wikimedia.org/w/index.php?title=User: Hoikka1&action=edit&redlink=1 (Creative Commons); pizza by Scott Bauer, United States Department of Agriculture, available at http://en.wikipedia.org/wiki/United_States_Department_of_Agriculture

Conclusions

Cities undoubtedly provide significant concentrations of the sorts of production and consumption process that lead to GHG emissions. Significant effort has been expended in seeking to understand and assess the particular contributions of specific cities to global GHG emissions, and, as the case study of Amman demonstrates, to assessing the potential for using such calculations as the basis for leveraging (international) sources of funding to support mitigation activities. Significant challenges to realizing these aims remain, both in terms of the practical issues of measuring and monitoring GHG emissions, and with respect to more fundamental issues about how they should be allocated. There is still significant uncertainty about how GHG emissions should be measured at the urban level, and whether this can be done in practice. For some, GHG emissions should continue to be allocated to the places in which they are produced, albeit that different indicators, for example per capita GHG emissions, may be more useful in terms of demonstrating the 'common but differentiated' responsibilities within and between cities. Others argue that only a consumption-based perspective, which illustrates how the consumption of goods and services leads to the production of GHG emissions, should be used.

Whichever approach is taken, what emerges from the analysis of how cities are contributing to climate change is a complex picture in which there are significant differences between total and per capita GHG emissions between countries, cities and urban communities. These differences are not only a result of existing patterns of, for example, economic production, urban morphology and the fabric of the built environment, but also reflect the historical development of urbanization and wider processes of economic production and consumption. Focusing on these complex geographies of GHG emissions at the urban level challenges the approach adopted by the process of international negotiation that responsibilities for addressing climate change can be allocated at the national level. It also raises questions as to whether adopting a single target and approach as the basis for developing urban climate-change strategies can sufficiently address the differentiated nature of contributions to GHG emissions within cities. Addressing these challenges will be a critical step towards creating more equitable, and more effective, urban climate-change responses.

Discussion points

- What are the advantages and disadvantages of both the production-based and consumption-based approaches to accounting for urban GHG emissions?

What are the implications for how we might allocate responsibility for addressing climate change at the city scale?
- Calculating carbon footprints: which aspects of your daily life do you think contribute most towards GHG emissions? First, either reflect on this by yourself or discuss in a small group. Then use one of the online tools suggested below to calculate your own carbon footprint. Were there any unexpected results? What did you overestimate, what did you underestimate, and why?
- What challenges do the issues raised in this chapter pose for setting GHG emissions targets at the urban level? How might these be resolved? Consider which sorts of indicator and target might be appropriate for different cities and for establishing different responsibilities within any one city. What might such principles mean in practice?

Further reading and resources

In addition to the core texts listed in Chapter 1, the following articles and reports are useful introductions to the debate on how to account for GHG emissions:

Dodman D. (2009) Blaming cities for climate change? An analysis of urban greenhouse gas emissions inventories, *Environment and Urbanization*, 21(1): 185–201.

Kennedy, C., Pinsetl, S. and Bunje, P. (2010) The study of urban metabolism and its applications to urban planning and design, *Environmental Pollution*, 159: 1965–73.

Hoornweg, D., Sugar, L. and Gomez, C. L. T. (2011) Cities and greenhouse gas emissions: moving forward, *Environment and Urbanization*, 23 (1): 207–27.

Online information about urban contributions to GHG emissions is available from the following sources:

- http://data.worldbank.org/topic/climate-change
- http://knowledge.allianz.com/?1384/green-cities-urbanization-carbon-footprint
- www.brookings.edu/reports/2008/05_carbon_footprint_sarzynski.aspx
- www.cdproject.net/cities.

Calculating your own carbon footprint can be done through the use of online tools. There are several available in different parts of the world, and these are just a selection:

- www.oneplanetvision.net/take-action/personal-action-plan-calculator/
- http://footprint.wwf.org.uk/
- http://carboncalculator.com/
- www.energysavingtrust.org.uk/Take-action/Your-carbon-footprint-explained.

4 Governing climate change in the city

Introduction

Climate change poses significant risks to cities (Chapter 2), and, at the same time, the concentration of GHG-producing activities in cities means that they have been seen as having increasing responsibility for the state of the climate (Chapter 3). How can and should cities respond to these challenges? What differences and similarities might be found between the vast array of cities, urban authorities and urban residents in responding to climate change? What are the consequent implications for urban development and for addressing climate change? In seeking to address these questions, researchers have begun to consider how climate change is being *governed* at the urban level. The aim of this chapter is to explore these different perspectives in order to provide a basis for the analysis of the urban mitigation and adaptation responses discussed in the following chapters.

In its most basic sense, governing involves intervening in order to direct or guide the actions of others. For some, governing is specifically associated with 'ruling' over others and is usually associated with the activity of government. For others, governing can also involve less obvious forms of intervention that seek to guide or influence the conduct of others. Both ways of thinking about governing are helpful for understanding responses to climate change, which have involved municipal authorities and the use of their powers to govern urban development and the day-to-day functioning of cities, as well as a range of other actors, who have, for example, sought to intervene to shape behaviour, create new forms of low-carbon energy or develop adaptive capacity.

These alternative approaches for understanding what governing involves also reveal some of the complexities of governing cities. To start with, cities are dynamic spaces within which multiple forms of rule and authority can be found. Municipal governments are one form of institution that creates rules that shape city life, but so do governments and their agencies operating at other scales, including regional government, national government and international organizations. Forms of rule-based authority also exist around particular sectors of urban life – whether that is vehicle standards for public-transport fleets, the levels of lighting required for public space, or building regulations for the construction industry. Sometimes, these rules and standards work in harmony between different levels of government or sectors of the urban economy, but sometimes they conflict. At the same time, other actors, such as utility companies, developers, architects, engineers, supermarkets and banks, have their own rules that shape their expectations of what 'normal' urban development is and, increasingly, of standards that are used on a voluntary basis to develop low-carbon or resilient infrastructure, products and services. As the following chapters will show, sometimes these voluntary approaches can go beyond existing formal 'rules', but they can also serve to channel urban development in particular directions that serve some interests rather than others. Among these more powerful actors are also organizations, including community groups, churches, environmental organizations and think tanks, that may lack formal power and resources but seek to develop new ways of acting in response to climate change that often rest on their access to knowledge and their ability to influence others to undertake collective action or to adopt new practices.

Some have argued that this diverse landscape of actors and approaches represents a break with the post-war era, when governments were the institutions most involved in governing the city. Others suggest there is more continuity, given that business and civil-society organizations have long had a role in urban development. In either case, what is clear is that it is no longer possible to suggest that it is *government* alone that governs – rather, the term governance has been adopted to signify the diverse range of government, business and civil-society organizations involved in governing (urban) society (Box 4.1). Recognizing that such actors operate across different scales (local, regional, national and international), and are also involved in creating new forms of partnership and collaboration that operate between and across these formal political arenas and serve to constitute new spheres of political authority, scholars have described the landscape within which urban responses to climate change are emerging as one of multilevel governance (Betsill and Bulkeley 2006; Gustavsson *et al.* 2009).

Understanding the multilevel-governance landscape within which the urban responses to the challenges of climate vulnerability and GHG-emissions contributions are emerging can provide a means of analysing how and why

> **BOX 4.1**
>
> **The new governance?**
>
> Although governance, in the sense of the multiple organizations involved in governing society, may not be new, it can be argued that, in the contemporary era, governance has led to a 'profound restructuring of the state' as it existed in the late twentieth century in many parts of the world, particularly developed economies, as is evident in:
>
> - a relative decline in the role of formal government in the management of social and economic relationships;
> - the involvement of non-governmental actors in a range of state functions at a variety of spatial levels;
> - a change from hierarchical forms of government structures to more flexible forms of partnership and networking;
> - a shift from provision by formal government structures to sharing of responsibilities and service provision between the state and civil society; and
> - the devolution and decentralization of formal governmental responsibilities to regional and local governments.
>
> Source: UN-Habitat 2009: 73

this is taking particular forms in different cities. This chapter explores these issues in two main parts. The first part outlines the history of urban climate-change governance, charting the initial involvement of municipal authorities and the approaches that were developed during the 1990s, as well as the changing dynamics of this movement over the past decade. The second part of the chapter considers the different modes of governing that have emerged in different cities and the institutional, political and sociotechnical drivers and barriers that are shaping urban responses. Whether it is through forms of rule or interventions to shape the conduct of others, governing is a complex and messy business. Interventions meet with resistance, produce unintended outcomes and can fail. This analysis provides the basis for considering in more detail, in Chapters 5 and 6, how mitigation and adaptation responses are being developed in a variety of different urban contexts. The chapter concludes with an overview of the main issues raised, as well as questions for discussion, further reading and resources.

Charting the emergence of urban climate-change responses

Urban responses to climate change date back over two decades, to the emergence of the issue as a matter of scientific and political concern. As we saw in Chapter 1, it was in the late 1980s that the connection between anthropogenic GHG emissions and the changing climate began to emerge on national and international political agendas. Mirroring this trend, some pioneering local governments also began to establish targets and timetables for taking action to reduce their contribution to GHG emissions. For example, following the 1988 international conference 'On the Changing Atmosphere' held in Toronto, in 1990 the city established a target of reducing carbon dioxide emissions by 20 per cent below 1988 levels by 2005 (Lambright *et al.* 1996; see also Bulkeley and Betsill 2003). In the UK and Germany, municipalities – including Leicester, Kirklees, Newcastle, Heidelberg, Munich and Frankfurt – started to develop climate-change policies as a result of their long history of interest in the issue of energy and the growing popularity of the notion of sustainable development and Local Agenda 21 (Bulkeley 2010; see also Collier 1997; Bulkeley and Kern 2006). Such targets and activities were largely symbolic, providing a stimulus to the emergence of a new agenda, rather than leading to immediate GHG-emissions reductions – indeed, most of these initial targets appear not to have been met.

Initially, these sorts of activity were confined to a small handful of municipal authorities. During the 1990s and 2000s, the number of municipalities engaged in addressing climate change grew significantly. At the same time, the nature of the urban response to climate change also began to shift, with the result that there is now a diverse set of urban authorities and other actors engaged in responding to climate change. In this history of the emergence and evolution of urban climate-change governance, a shift from an approach largely based on *municipal voluntarism* to one focused on *strategic urbanism* can be traced.

Municipal voluntarism

As outlined above, initial efforts to address climate change at the city scale were largely voluntary in nature and focused on mitigation. Individual municipalities, primarily in North America and Europe, declared their intentions to reduce GHG emissions and set about establishing strategies and measures through which this could be accomplished. Many of the municipalities involved had previously been engaged in issues of urban sustainability and regarded acting on climate change as in line with their commitments in this field. At the same time, given the small contribution each could make to the challenge of reducing GHG emissions, these pioneering municipal authorities recognized the need for some level of collective response. In the early 1990s, municipalities began to organize projects and networks that would enable them to connect with one another, share information about the challenges and opportunities of responding to climate change, and mobilize politically on the issue.

An important feature of these early efforts at a collective urban response is that they were *transnational* in scope – that is, that they explicitly sought to connect municipal authorities that were responding to climate change in different countries. Municipal authorities have a long history of forming such associations and networks – for example, the International Union of Local Authorities was founded in 1913, and the Council of European Municipalities and Regions (CEMR) was established in 1951. However, it was not until the early 1990s that such transnational municipal networks began to take on a distinctively environmental agenda. During the early 1990s, three transnational municipal networks focusing on climate change were established. Although they had similar goals, their diverse origins point to the multiple ways in which climate change was becoming an issue on urban agendas during this period.

In 1991, Local Governments for Sustainability (ICLEI) (then known as the International Council for Local Environmental Initiatives), a Toronto-based local government association, formed the Urban CO_2 Reduction Project, which was designed 'to develop comprehensive local strategies to reduce greenhouse gas emissions and quantification methods to support such strategies' (ICLEI 1997). Funded by the US Environmental Protection Agency (EPA), the City of Toronto and several private foundations, fourteen municipalities in Europe and North America took part in this initial pilot project (Bulkeley and Betsill 2003). In 1993, ICLEI launched the CCP programme at a meeting in New York attended by 150 municipal leaders, and, later in the same year, more than eighty representatives of cities in twenty-three European countries attended a meeting in Amsterdam to launch the CCP-Europe campaign. The second network to be formed at this time, the Climate Alliance, was established in 1990. Based in Frankfurt, the network sought to encourage solidarity between municipalities in Germany and its neighbouring, German-speaking countries and indigenous people in the Amazon region. Addressing climate change, it was argued, meant reduction in GHG emissions in industrialized countries such as Germany, together with work to protect indigenous communities and the rainforests within which they lived.

In keeping with this framing of the climate-change problem, the aim of Climate Alliance is to reduce municipal GHG emissions to 50 per cent below 1990 levels by 2030 and protect the rainforest through partnerships and projects with indigenous rainforest peoples (Kern and Bulkeley 2009). A third network was initially formed in 1990 by six municipalities, including Besançon (France), where it is headquartered, Newcastle (UK) and Mannheim (Germany). These municipalities were involved in a European Union-funded project, Energie Cités, and the network that emerged from this initial collaboration was formally constituted as an association of municipalities in 1994, with sixteen members. Although addressing climate change was central to the aims of Energie Cités, the network established a remit specifically to address all aspects of energy use and generation at the municipal scale.

In Europe, membership of climate-change networks grew steadily during the 1990s, although it had reached a plateau by the end of the decade (Kern and Bulkeley 2009). An assessment conducted in the early 2000s showed that, over the decade since they had been initiated, 'almost 1,400 European cities and towns, including many capital cities, such as Amsterdam, Rome, Stockholm and Berlin' had joined at least one of these networks (Kern and Bulkeley 2009: 316). With over 1,000 members, Climate Alliance was the largest network, while Energie Cités had over 160 individual members in twenty-five European countries, and ICLEI Europe had some 120 members in sixteen countries. Towards the end of the 1990s, new regional campaigns by ICLEI and its CCP programme saw membership increase in Australia and the US, as well as in Asia and Latin America. Despite the growth in membership, however, reported actions were primarily focused on the reduction of GHG emissions from within the municipality and particularly concerned with issues of energy efficiency.

In effect, this first wave of urban responses was typified by what might be termed *municipal voluntarism* – a focus on the voluntary activities of municipal authorities as a means of building capacity to address climate change. Although targets and timetables for reducing GHG emissions across urban communities were often adopted during this period, municipalities were faced with challenges of how to measure GHG emissions (Chapter 3), the limited nature of their policy- and plan-making powers in relation to critical areas of infrastructure and urban development, a lack of capacity to implement those measures, and political challenges concerning how and why addressing climate change should be regarded as more important than other urban issues, such as wealth creation or increased mobility (Bai 2007; Betsill and Bulkeley 2007; Bulkeley *et al.* 2009; Romero-Lankao 2007; Schreurs 2008).

As municipalities sought to address these challenges and build the capacity to act, climate change was often reframed as an issue that could achieve multiple goals or realize co-benefits – including air pollution, health, congestion, energy security and so on. Nonetheless, many municipalities struggled to find the means through which to engage broader communities and the range of relevant stakeholders in taking action on climate change in the city. Furthermore, beyond the emerging forms of transnational co-operation discussed above, municipal authorities were often left to fend for themselves in this issue arena, with little attention being paid to the urban dimension of responding to climate change by either national governments or the emerging international regime (for an exception, see Sugiyama and Takeuchi 2008; Granberg and Elander 2007). Lacking support from other levels and arenas of governance, and without sufficient resources to act, some of the initial promise of an urban response to climate change began to fade. Although the ideal of the municipal voluntarism that characterized urban climate governance during this decade was based on an integrated, evidence-based approach to climate planning and policy, the challenges of institutional capacity and of political economy that were encountered as authorities sought to engage in responding to climate change beyond their own operations led to a more piecemeal and opportunistic approach. Although some cities were able to develop sufficient capacity and political will to overcome such barriers, many witnessed a growing gap between the rhetoric of a need for an urgent response and the realities of governing climate change on the ground (Bulkeley 2012).

Towards strategic urbanism?

By the end of the 1990s, membership of the three transnational municipal networks – ICLEI CCP, Climate Alliance and Energie Cités – had begun to stabilize. In the light of growing international uncertainty about the fate of the Kyoto Protocol, municipalities, along with national governments, appeared to be more reluctant to commit to taking action. However, just as it seemed that the interest in, and momentum behind, urban responses to climate change might be lagging, a second wave of municipal engagement with climate change began. Driven by renewed national and international commitment to address the issue, as well as increasing scientific evidence of the scale and severity of the climate-change problem, existing municipal networks began to expand their membership, and new associations also began to form. This was not simply a matter of international targets leading to national action and pressure on municipalities to act. Rather, momentum emerged from the opportunities, contradictions and tensions across this multilevel-governance system.

In seeking to build and expand their membership, existing transnational municipal networks began to organize regional or national campaigns: for example, Energie Cités established a regional programme in Poland, and ICLEI established CCP programmes in South Asia (Box 4.2), South East Asia and Latin America. At the same time, a lack of political action at the national level also served as an opportunity for the emergence of municipal action on climate change. In the US, the growing intransigence of the George Bush administration with regard to climate change led some municipal governments to form the US Mayors' Agreement (Gore and Robinson 2009). Although initiated in 2000, it was in 2005 that the call by the mayor of Seattle, Greg Nickels, for mayors across the US to take action on the issue led to action (Gore and Robinson 2009: 142). Following an initial agreement among ten of the leading US cities on climate change, a further call to action attracted over 180 mayors, and, by 2009, over 900 mayors had signed up to the Climate Protection Agreement (Gore and Robinson 2009: 143). In a parallel initiative, the Sierra Club launched the Cool Cities campaign in 2005 and now has over 1,000 members. Likewise, in Australia, during the early 2000s, the Howard government's position that Australia should not sign the Kyoto Protocol served to spur many local authorities to join the CCP programme in Australia and to engage their political representatives in recognizing the importance of the issue. This move to engage local politicians with the climate-change agenda has been replicated globally, most recently with the launch in 2009 of the European Covenant of Mayors, which requires signatories to pledge to go beyond the EU target of reducing carbon dioxide emissions by 20 per cent by 2020 through the formation and implementation of a sustainable energy action plan (Covenant of Mayors 2011a) and which, in 2011, had more than 2,000 members (Covenant of Mayors 2011b).

The formation of additional transnational municipal networks focused on specific *types* of city has also been important in the past decade. Of these, the highest-profile example is that of the C40 Cities Climate Leadership Group, which now involves forty of the world's largest cities as full members and a further eighteen cities as affiliate members (Figure 4.1). Instigated by the then mayor of London, Ken Livingstone, and his deputy, Nicky Gavron, together with the Climate Group, a not-for-profit organization based in London, the original network was formed by eighteen cities in 2005 as a parallel initiative to the Group of Eight (G8) Gleneagles summit on climate change. In 2007, this network entered into a partnership with the Clinton Climate Initiative and expanded its membership to include forty of the largest cities in the world. In a similar fashion to the CCP, Climate Alliance and Energy Cités networks that have gone before it, C40 sets considerable store on sharing knowledge and experience, showcasing best practices and providing resources for its members. It also seeks to make strategic claims for the importance of cities in addressing climate change, and in turn to use this positioning to gain political leverage of resources within particular national contexts (Hodson and Marvin

BOX 4.2

ICLEI South Asia

ICLEI Local Governments for Sustainability was founded in 1990 as a network association of local governments. Its five regional chapters (one in each continent) provide training, consulting and overall support for local governments to tackle environmental problems. The organization has a membership of over 1,200 local and regional governments from over seventy countries.

ICLEI's activities in South Asia started in 2001 with the CCP programme, an initiative first brought to India with the support of USAID. The programme's objective was to kick-start local mitigation and adaptation actions. It quickly evolved into a series of climate-change initiatives and projects, such as the 2005–10 Local Renewables Project, an initiative to promote the adoption of renewable energy and energy efficiency in cities. ICLEI, in partnership with local governments, international donors and other stakeholders, has prepared GHG-emissions inventories for over fifty South Asian cities, established pilot mitigation and adaptation projects, and developed guidelines on 'urban low-carbon actions'.

Through these and other projects, ICLEI South Asia has played a key role in raising awareness of climate-change issues in local governments across India. The international ICLEI network has provided local officials with the opportunity to learn about and discuss the relevance of climate change. It has also offered opportunities to respond to the challenges via specific pilot projects. For example, as a result of these partnerships, the Indian cities of Bhubaneswar, Coimbatore and Nagpur developed renewable-energy and energy-efficiency policies and strategies. This experience was instrumental in the final formulation of India's Solar Cities Programme, launched in 2008 by India's Ministry of New and Renewable Energy, where ICLEI South Asia is one of the key delivery partners.

The work of ICLEI also shows how the response of Indian cities to climate change often takes the form of energy interventions. Cities working with ICLEI have engaged actively in mitigation programmes, in part attracted by the financial savings resulting from renewable-energy and energy-efficiency measures at the municipal level. In this way, energy issues have become an entry point for cities to start engaging with climate change.

For more information see: www.iclei.org/sa.

Andrés Luque, Department of Geography,
Durham University, UK

2009, 2010). Unlike the approach taken by transnational municipal networks in the 1990s, C40 specifically seeks to influence international and national agendas through various forms of advocacy, including disseminating best urban practice, hosting events and showcasing the contribution that global cities can make to addressing the problem (Arup 2011b; Carbon Disclosure Project 2011). Working with the Clinton Climate Initiative, the network has mobilized finance, knowledge and additional partners to undertake specific projects, including, for example the Energy Efficiency Building Retrofit programme, which seeks to develop new models of financing energy savings in commercial buildings, and Climate Positive, a project undertaken in collaboration with the global architecture and urban development firm Arup, which seeks to bring together demonstration projects of low-carbon development in cities worldwide.

The emergence of new networks and the extension of ICLEI's CCP programme have led to a growing engagement with issues of climate change in cities in the Global South. For example, some fifty local authorities in India have participated in ICLEI South Asia's Roadmap project to conduct emissions inventories and develop climate-change action plans (Box 4.2), and twenty of the forty cities included in the C40 Cities Climate Leadership group are located in countries in the Global South. At the same time, mainstream development organizations have begun to consider the issue – most notably the World Bank, which released its report *Cities and Climate Change; an urgent agenda* in 2010, and UN-Habitat, whose *Global Report on Human Settlements 2011* focused on cities and climate change. Coupled with this increasing interest among development organizations and cities in the Global South has been a growing acknowledgement that issues of climate-change vulnerability and adaptation also need to be considered at the urban level. Existing networks, most notably ICLEI, have begun to focus on climate adaptation and are seeking to engage cities through the concept of resilience, epitomized in the annual conference on Resilient Cities first held in 2010, and through new toolkits, strategies and programmes focusing on adaptation across their regional offices. In addition, the Rockerfeller Foundation has established the Asian Cities Climate Change Resilience Network, a network of ten cities explicitly focused on climate adaptation, and UN-Habitat is working with cities in Asia, Africa and Latin America through its Cities and Climate Change Initiative to support urban-adaptation responses. This is unusual, not only in the urban context, given the predominant focus on mitigation, but also in terms of climate governance more broadly, where most transnational governance initiatives have also been focused on mitigation. However, outside these initiatives, research suggests that issues of adaptation remain marginal. For the most part,

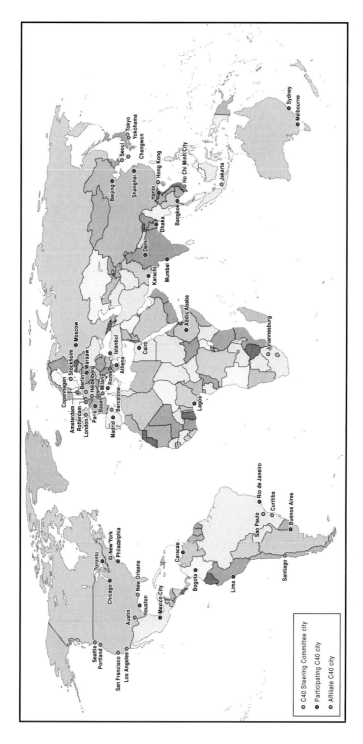

Figure 4.1 C40 Member and affiliate cities 2011

> urban governments in low- and middle-income nations have not considered adaptation seriously. For instance, in India, Chile, Argentina and Mexico, central government is beginning to take an interest in adaptation, but this interest has yet to engage the interests of the larger, more powerful national ministries or agencies or city and municipal governments.
>
> (Satterthwaite 2008a: 14)

An analysis of urban climate-change strategies in ten cities in India, China, Mexico, Brazil, Australia, South Korea, Indonesia and South Africa found that mitigation remains the focus of urban climate policy, despite the arguably more pressing adaptation issues facing cities in the Global South (Bulkeley *et al.* 2009).

Alongside this growing diversity of cities engaged with the climate-change agenda has been the growth of interest in urban climate governance among private and civil-society actors. Networks such as C40 and ICLEI increasingly work with private-sector actors – including property developers, financial organizations, architects and so on – and there is also evidence that such actors are establishing their own networks concerned with issues of climate-change mitigation and adaptation. For example, from 2006 to 2011, HSBC established the Climate Partnership, one part of which is concerned with developing urban responses to mitigation in its global centres – New York, London, Hong Kong, Mumbai and Shanghai. Property and infrastructure development organization Aecom also runs a Global Cities Institute, which brings together its staff to address critical urban-sustainability issues, including climate change, in one city each year. In contrast, addressing climate change in the city is also an issue that is being championed by grass-roots organizations concerned with issues of justice and achieving a transition to a post-oil future, such as the Transition Towns movement in the UK (Chapter 7). Although the motivations and nature of urban climate governance emerging through these diverse engagements are significantly different, together they signify the common trend for urban climate governance to be undertaken by a range of non-state actors in the city.

Evaluating urban responses to climate change

Over the past decade, the urban governance of climate change has therefore been transformed. From being the preserve of a few pioneering municipalities and a matter primarily for cities in North America, Europe and Australia, efforts to address climate change in the city can now be found across different global regions, include both large global cities and those that may be particularly vulnerable to the effects of climate change, engage large private-sector actors and different forms of social movement, and have begun to take a more central place in urban political agendas. The result is a form of climate governance that we might call *strategic urbanism* – signifying the growing alignment between addressing climate change and core municipal concerns, and the more direct, political approach that municipal authorities and other urban actors have begun to take to the issue (Hodson and Marvin 2010). In part, this new urban response to the challenge of climate governance can be explained by a growing sense of frustration at the pace and nature of the international negotiations (Hoffmann 2011). However, it is also driven by distinctly urban concerns. For some cities, these concerns relate to the opportunities that are being created by the emerging carbon market and the constraints that are being imposed on GHG emissions by the pursuit of national targets. In this vein, While *et al.* (2010: 82) suggest that processes of eco-state restructuring are now focused on 'carbon control', creating a 'distinctive political economy associated with climate mitigation in which discourses of climate change both open up, and necessitate an extension of, state intervention in the spheres of production and consumption'. Mike Hodson and Simon Marvin make a similar argument for the growing strategic importance of climate change in terms of the need for cities to secure resources for economic development, particularly in the forms of energy and water, which in turn is leading to a distinct set of urban political economies orchestrated by and through the climate-change issue (Hodson and Marvin 2010). As they also argue, however, this is an agenda underpinned by concerns for *resilience* alongside security, where municipalities, and an increasing array of other urban actors, are concerned to ensure that they are able to adapt to the impacts of climate change.

The result is a complex landscape of urban climate governance (see the city case study on Melbourne). Although the voluntary, municipal approaches that characterized the early responses to climate change in cities persist, new forms of strategic urbanism are also emerging. A plethora of networks and associations of cities that cut across regional and national boundaries have emerged, and partnerships between government, corporate and civil-society actors are now a common way in which climate change is being addressed. Among all of this activity, the total number of cities that have committed to act on climate change is difficult to determine. The CCP and Climate Alliance have membership of over 1,000 cities, and the European CoM has over 3,000 members, but there is some degree of overlap. Beyond such transnational networks, however, there are nationally and regionally based alliances that have several hundred cities as members, such as the US Mayors Climate Protection Agreement and the UK's Nottingham Declaration on Climate Change, and individual cities have also pledged their own commitments. At the same time, climate change has, whether through national policy or international obligations, become part of the ways in which different urban policy and economic sectors conduct their daily business, whether this is in terms of EU standards for buildings, codes for the insurance industry or the growing importance of the carbon market for the financial sector.

Our understanding of how and why climate change has become a critical part of urban policy agendas has for the most part been driven by case-study research (Melbourne case study). However, in the US, large-scale assessments have been made of a range of contextual variables that might explain the reasons why municipalities have undertaken climate policy or planning. Some authors have found that a combination of vulnerability to climate change, low levels of contribution to the climate-change problem in terms of GHG emissions or economic activities that consume high levels of energy, and 'civic capacity' (indicated by, for example, income, levels of education, political support) can explain the likelihood of membership of the CCP campaign (Zahran *et al.* 2008; Brody *et al.* 2008). However, other US-based studies have corroborated earlier case study research and have found that it is the political/institutional support within municipalities that most clearly explains the adoption of climate-change policies (Pitt 2010: 867) and, to some extent, the level of action being undertaken (Krause 2011: 58).

Assessing how and with what implications such commitments have led to action to mitigate and adapt to climate change is discussed in detail in the following chapters. Assessing the impacts of urban responses to climate change can be done in several ways. First, and perhaps most obviously, is the question of the direct effect that such strategies and measures have had in terms of reducing GHG emissions and reducing vulnerability. In many ways, this is the hardest form of evaluation to undertake, as it relies on consistent ways of measuring the impacts of a whole array of interventions, and on being able to link them to specific outcomes. This is one of the goals of the measurement approaches that were discussed in Chapter 3, but these remain very challenging to implement in practice, and there is, to date, limited evidence of the extent to which specific schemes have achieved GHG-emissions reductions or policy goals have been met. A second set of effects that urban responses to climate change may have is in reframing core urban concerns and processes – e.g. development, service provision, mobility and public space – in such a manner that climate change becomes central to how they are envisaged and implemented. The analytical challenge here is that, although the rhetoric of many urban actors, from municipal governments to banks, churches to community groups, bears the hallmarks of addressing climate change, in practice the extent to which it has become integrated into business as usual and has served to make a difference to how this is conducted is often difficult to discern. A third way in which urban responses to climate change might be

CITY CASE STUDY

Governing climate change in Melbourne

Melbourne is the second largest metropolitan region in Australia, covering approximately 8,000 km² and home to an estimated population of close to 4 million people. The capital of the Australian state of Victoria, the Greater Melbourne region comprises thirty-one local authorities, many of which have been pioneers of urban responses to climate change.

Over the past two decades, policy responses to the issue of climate change in Australia have waxed and waned. Initial enthusiasm in the early 1990s, when the federal government was one of the first nation-states to commit to the UNFCCC, soon gave way to more muted commitments to action and to outright opposition to the formation of international agreements and binding emissions-reduction targets (Bulkeley 2001). During the late 1990s and early 2000s, Australia remained opposed to an international climate treaty and continued to argue that undertaking substantial reductions in GHG emissions would place an unfair economic burden on its export industries and labour force. However, despite this international position, over the same period the Australian federal government developed several strategies and measures for addressing GHG emissions domestically, primarily based on an argument that such actions could be undertaken on a 'no-regrets' basis, because they would yield economic savings. Among the policy measures adopted was the development of ICLEI's CCP progamme, aimed at encouraging local governments across Australia to engage with climate-change mitigation (Bulkeley 2000). During the 2000s, the CCP Australia programme grew substantially, funded by federal, state and local governments, so that, by 2008, some 234 Australian councils participated in the programme, 'representing about 84% of the Australian population' (ICLEI Australia 2009).

Urban responses to climate change in Australia have been developed within this multilevel context of a national government that is hostile to international commitment but that has supported the development of a transnational municipal network across Australia. This multilevel context is further complicated by the actions of particular states. During the 1990s, the Victorian government, a Liberal Party administration under the premiership of Kennet, undertook various reforms to privatize the energy sector and to promote economic growth, with little regard for environmental consequences (Jessup and Mercer 2001). Although climate change was acknowledged as a policy issue with the publication, in 1998, of the *Victoria's Greenhouse Action: Responding to a Global Warming* strategy, it was regarded by some as 'little more than a public relations' exercise' (Jessup and Mercer 2001: 23). After 1999, the Labour administration developed a more comprehensive approach to climate policy. The 2002 Victorian Greenhouse Strategy set out a range of measures to encourage the development and use of renewable energy and reduce demand for energy, including the development of energy-efficiency standards for buildings so that new developments were required to attain a 5★ rating from 2005, the promotion of GreenPower energy, support for the CCP programme in regional and rural Australia, and the formation of regional partnerships between local governments to pool efforts and resources in addressing climate change.

The development of climate-change responses across Melbourne bears the hallmarks of these federal- and state-government responses. Although some municipalities across Melbourne had been involved in issues of energy efficiency and renewable energy during the 1980s and 1990s, it was the foundation of the CCP programme in Australia in 1998 that gave impetus to the direct involvement of local authorities in the city with the climate-change agenda. During the period 1998–2002, municipalities across Melbourne joined the CCP programme and began to adopt its milestone approach for addressing in-house GHG emissions. In the wake of the 2002 Victorian Greenhouse Strategy, and with the additional support offered by the Victorian state, municipalities that had been part of the CCP progamme started to form partnerships in order to progress their internal policy agendas. One such partnership was the Northern Alliance for Greenhouse Action (NAGA), whose population comprises some 25 per cent of the population of Victoria (NAGA 2008). NAGA was established as an informal network for sharing information and developing new projects among six of these pioneering authorities. Having completed the 'milestones' involved in the CCP programme, by the mid 2000s municipalities in the north of metropolitan Melbourne, and in particular those that had adopted the CCP programme early on, began to develop more ambitious targets and innovative approaches (Table 4.1). In 2002, the City of Melbourne adopted a target of reaching 'zero net emissions' by 2020, followed in 2007 by Moreland. Despite the recent recognition by the City of Melbourne in its update of the 2002 strategy that the target of reaching 'zero net' emissions will not be realized, the policy ambition to achieve significant cuts in GHG emissions has been reiterated and appears to be spreading across the NAGA councils.

Seeking to explain the foundations of this success, participants suggest that it is the importance of showing leadership in this area that is the primary driver behind their achievements:

> We couldn't show how much money we have saved in total through all of these things. We probably can't show, I'm probably being unfair but we probably couldn't show how much it cost us either. We know as an organization that we've gained reputation . . . and we know that we are making Melbourne . . . a better place to be a competitive 21st century city.
>
> (Interviewee, Melbourne, July 2008)

The strategic importance of showing leadership on the issue of climate change is one signal that the largely voluntary and municipally orientated approaches developed in the late 1990s in Greater Melbourne and other municipalities have been replaced by a more strategic vision. In the case of the City of Melbourne, this leadership has been demonstrated through membership of the C40 network. In the 2008 'Zero Net Emissions by 2020 Update', there is 'growing recognition that the City of Melbourne needs to align with other like-minded climate change cities' globally (City of Melbourne 2008: 13). This involvement with an international network provides access to information and resources and allows the City of Melbourne to demonstrate its strategic importance in relation to addressing climate change. This is not, however, without challenge (Bulkeley and Betsill 2011). Despite the interest and enthusiasm of some policy entrepreneurs, responding to climate change still means working within the framework of municipal governance, where resources remain limited and climate change remains a marginal agenda for many of the day-to-day workings of the municipality. A second challenge related to the conflict between environmental and economic agendas, an issue found to be particularly pressing at the urban fringe, where imperatives for economic growth and development pressures are strong. A final challenge concerned the feasibility and delivery of ambitious targets and the need to avoid the creation of goals simply being conceived for political ends, with little prospect of being fulfilled. The dilemmas of setting realistic targets, managing expectations and still seeming to 'lead' the field were evident in the discrepancy between some policy rhetoric concerning the importance of local action on climate change, the continued focus on internal emissions reductions for many councils, and the high and rising levels of emissions across the metropolitan area. Although new forms of urban climate governance are emerging, many 'old' political issues remain at the core of the challenge of governing climate change in cities (Bulkeley and Betsill 2011).

Table 4.1 The emergence of climate-change governance in Melbourne

Milestone	Goal	Approach
1997/1998 Darebin, Manningham, Melbourne and Moreland join Australia CCP	To reduce emissions within the council and/or community of the order of 20–30% by 2010	CCP Milestone programme with financial assistance from federal government
1999–2002 Banyule, Hulme, Nillumbuik, Whittlesea and Yarra join CCP Australia	To reduce emissions within the council and/or community of the order of 20–30% by 2010	CCP Milestone programme with financial assistance from federal government
2001 Moreland Energy Foundation Ltd founded	To reduce GHG emissions across Moreland	A not-for-profit company established by Moreland City Council after the sale of the Brunswick Electricity Company. With support from the council, it runs a series of community programmes addressing GHG-emissions reductions and other issues of energy poverty and security
2002 Darebin Resource Efficiency Fund established	To provide funding for energy-efficient schemes within the council	A fund designed for investments in council buildings and facilities
2002 Community Power scheme established	To provide access to green power products for households	A partnership between Darebin, Whitehorse and Moreland City Councils, the Moreland Energy Foundation and energy providers, to sell accredited green power products together with information and advice to local households
2002 Victorian Greenhouse Strategy	To develop a framework for addressing climate change across the state; to work with local governments to reduce emissions of GHGs	Among many other measures, establishes funding for co-ordinators of regional partnerships of local authorities in Victoria
2002 NAGA established	To achieve significant greenhouse abatement by delivering effective programmes and leveraging council, community and business action	Established as an informal network between six northern municipalities in metropolitan Melbourne
2002 City of Melbourne Zero Net Emissions by 2020 Strategy	Aims to reduce the contribution of the City of Melbourne to GHG emissions by 2020	Combination of measures focused on energy efficiency in the built environment, renewable-energy supplies and carbon offsetting

continued . . .

Table 4.1 ... *continued*

Milestone	Goal	Approach
2005 Victorian Greenhouse Strategy Action Plan Update	To consolidate and extend climate-change policy in the state, including expanding the regional partnership initiative	Provided funding for NAGA and is expansion to the current membership
2007 Moreland Climate Action Plan 2007–12	Adopts a zero net emissions target by 2020 for the council and 2030 for the community	Combination of measures including engaging the community in reducing emissions, switching to renewable-energy sources and offsetting
2007 Commission for Melbourne Climate Change Taskforce	To assess the impacts of climate change on Melbourne, the potential for mitigation, and opportunities for Melbourne	A coalition of eighty private and public actors commissioned and reviewed research, leading to the publication of *FutureMap* in 2008. The Taskforce has since developed a range of specific initiatives (e.g. green roofs competition, staff travel policy) and action groups (e.g. on retrofitting, clean-coal technologies)
2008 City of Melbourne Zero Net Emissions by 2020 Strategy Update	Aim to achieve zero net emissions for the council by 2020, and a reduction of 50–60% across the community	Combination of measures focusing on the commercial and residential sectors, passenger transport and decarbonizing energy supply
2008 Towards Zero Net Emissions project	To quantify regional emissions and assess the options for achieving zero net emissions	Research project sponsored by the Victorian Sustainability Fund and undertaken by Arup Ltd, due to be completed in autumn 2009
2008 Moreland Solar City initiative	To develop a sustainable community focused on reducing emissions of GHG and the development of renewable energy	A partnership between MEFL, Moreland City Council, the Brotherhood of St Laurence, Sustainability Victoria to retrofit low-income homes, redevelop a brownfield site, engage communities in reducing energy use, and establish an energy services company. One of seven solar cities projects sponsored by the federal government solar cities programme, it received AU$4.9 million in funding

Source: Adapted from Bulkeley and Schroeder 2009

Figure 4.2 Melbourne: climate governance in action

Source: Cloud 9 Aerial Photography for live.org.au

considered to have had an effect is through their influence on other levels of government decision-making and on the actions of other actors beyond the urban sphere. Here it is possible to discern how, since the 2007 Conference of the Parties in Bali, where local governments were the second largest delegation, they have made their presence felt at successive international negotiations, and the growing recognition of urban climate-change concerns on national agendas. It is also notable that leading corporate actors in the climate-change arena, including HSBC, Arup and Cisco, have chosen to focus on cities as a means through which to address their own concerns in this arena. This suggests that cities are now firmly on the climate-governance map.

Understanding the nature of urban climate governance

The emergence, development and consolidation of urban climate-change governance over the past two decades, as outlined above, reflects a diverse range of drivers and political circumstances. Understanding how municipal governments and other urban actors have sought to act on the intentions expressed through networks, targets and strategies means considering how climate-change governance has evolved in practice in the context of specific urban places. Given the changing landscape of urban governance, as described in the opening section of this chapter, it is no surprise that the means that have been sought to govern climate change within cities have not been confined to traditional instruments of municipalities, such as land-use zoning, but have encompassed a broad range of direct and indirect forms of intervention. Despite this variety, distinct *modes of governing* can be identified that have been deployed by municipalities and other urban actors in order to address climate change. As set out below, each of these modes relies on different sorts of policy instrument and intervention, as well as on the mobilization of different forms of resource and power. As the process of governing climate change in the city has gathered pace, a number of drivers and barriers have been encountered, including issues of *institutional capacity*, of *political economy* and those relating to the possibilities for *sociotechnical* transition. Although it is possible to identify drivers and barriers that can be found across different urban contexts, as explored below, their exact nature will vary with specific urban circumstances. In seeking to explain the nature of urban climate governance, the remainder of this section sets out in general terms the modes of governance that have been used, and the drivers and barriers that have been encountered, in order to establish the basis for considering how mitigation (Chapter 5) and adaptation (Chapter 6) have been pursued.

Modes of governing climate change in the city

In seeking to develop urban climate-change governance, municipal authorities deploy a number of distinct modes, or ways, of governing (Alber and Kern 2008; Bulkeley and Kern 2006; Bulkeley *et al.* 2009). In essence, a mode of governing refers to a specific set of processes and techniques through which governing is pursued. Such processes can, in turn, be regarded as reliant on different forms of governing capacity or power, as encompassing tools, technologies and artefacts, and as based on different rationalities about the nature and purpose of governing climate change (Bulkeley *et al.* 2007). In relation to climate change, research suggests that municipal authorities deploy four distinct modes of governing (Table 4.2). Although each mode is underpinned by different sets of processes, logics and techniques, they are not mutually exclusive; rather, municipalities tend to deploy a combination of these modes at any one time.

First, *self-governing*, the means through which municipalities seek to govern their own operations. This can involve various approaches intended to manage municipal operations (buildings, fleets etc.), to procure more sustainable products (such as purchasing renewable energy), or to establish demonstration projects as ways of leading by example. This mode of governing has, at least in part, been facilitated by new forms of organizational management within municipalities, sometimes termed new public management, which have included target setting, new financial instruments and performance contracting (Bulkeley and Kern 2006). In terms of urban responses to climate change, this mode of self-governing has been dominant in cities in both the North and Global South. As Gore *et al.* (2009: 10) suggest, the reasons that such approaches are dominant is not surprising, for

> they require minimal or no community buy in, creating little political debate; they usually produce direct returns with respect to cost savings; they produce quick, verifiable reductions in emissions; and actions to reduce [local government] emissions are promoted as important first steps by ICLEI and the Federation of Canadian Municipalities.
> (Gore *et al.* 2009: 10)

These benefits are, of course, to be welcomed, but the strong focus on self-governing responses is also worth a note of caution. Given that the proportion of GHG emissions or of vulnerable infrastructures and communities that lie directly within municipal control is small in comparison with the city as a whole, an undue emphasis on this mode of governing may create an overly optimistic approach to the potential for addressing climate change in the city, while at the same time serving to address only a small piece of the overall urban climate-change puzzle.

A second mode of governing, *provision*, involves the development of low-carbon or resilient infrastructure systems, as well as the delivery of services and goods that have a lower-carbon footprint or seek to improve adaptive capacity. This mode of governing is important, for it has the potential to have significant implications for the practices of production and consumption, through which GHG emissions and vulnerabilities are created. However, although municipalities in developed economies traditionally had significant influence over infrastructure systems and the services they provided – such as energy, water and waste – as the Introduction to this chapter made clear, the past few decades have witnessed the 'splintering' of such systems of provision. In cities in the Global South, municipalities often have some level of responsibility for such services, but they remain partial in their extent and coverage, and the challenges of meeting basic needs for access and affordability mean that the scope to consider issues of climate-change mitigation and adaptation is frequently limited (UN-Habitat 2011; Satterthwaite 2008a). Intervening through provision requires municipalities and other actors to develop new ways of transforming the infrastructures and systems through which such services are provided and needs are met. Such a mode of governing can be undertaken through demonstration projects and the introduction of new forms of energy, water and waste technologies into existing systems. However, wider transformations of such systems typically require, not only small-scale innovations, but the realignment of a range of actors, resources, technologies and social practices (Geels 2002; Smith *et al.* 2010). Despite these limitations, municipalities are increasingly deploying new forms of provision in relation to mitigation (Chapter 5) and adaptation (Chapter 6) and, with other actors, creating new ways of experimenting with climate change in the city (Chapter 7).

As a mode of governing, *regulation* – approaches intended to oversee, guide and determine particular conditions, ways of operating, behaviour and standards – are perhaps the least used in the climate-change domain (Bulkeley *et al.* 2009; UN-Habitat 2011). Nonetheless, it is possible to identify three sets of mechanisms that municipalities have used in this mode – financial regulations (e.g. taxation, subsidies), for example related to congestion charging, land-use planning (e.g. requirements for mixed-use, the density of development, zoning and the inclusion of renewable energy), and setting standards, particularly in relation to new and refurbished buildings. This mode of governing is reliant on the ability of municipalities to enforce regulation and to sanction those that do not comply. On the one hand, this means that

Table 4.2 Modes of governing climate change in the city

Mode of governing	Policies and mechanisms	Logics and capacities required	Advantages	Limitations
Municipal self-governing	Management of local authority estate; procurement; demonstration	New public management	Under the direct control of the municipality and can provide quick, measurable and cost-effective action. Can be used to demonstrate leadership and commitment	Such measures can only address a small proportion of urban GHG emissions or vulnerable locations/communities. They may be limited to those that can provide a financial return within the (short) time horizons of local governments
Municipal provision	Low-carbon and resilient municipal infrastructure systems, goods and services	Assembling and aligning diverse social and technical entities	Has potential for addressing significant sources of emissions and vulnerability, and could also improve access and affordability of services	Municipal capacity is hampered by a lack of finances, dependency on the terms and conditions of capital loans, and a limited remit for providing energy, water, waste and transport. In contexts where there is a lack of basic services, developing climate-change responses is unlikely to be a priority
Municipal regulation	Financial instruments (e.g. taxes, subsidies); land-use planning, codes, standards etc.	Traditional forms of authority including enforcement and sanction	Regulative measures can provide the basis for transparent and effective policy. They may also yield additional revenue, which can be invested in additional low-carbon measures	Regulative measures can be difficult to implement because of concerns about their impact on businesses or particular sections of the community. Regulations are difficult to apply retrospectively (e.g. to existing buildings), and governments are often reluctant to regulate individual behaviour, meaning that the application of such measures may be confined to a small proportion of total urban GHG emissions. In a context

			of limited municipal capacity, regulations can be difficult to monitor and enforce
Municipal enabling	Information and awareness raising; incentives; partnerships	Diverse modalities of power, including inducement, persuasion and seduction (Allen 2004)	Enabling measures can require relatively little financial or political investment. They enable municipal governments to benefit from the resources and capacities of a range of other urban actors in reducing GHG emissions. Through involving a range of different partners they may increase the democratic mandate for acting on climate change	Enabling measures are dependent on the goodwill and voluntary actions of businesses and communities, which may not be forthcoming. Assessing and verifying the impact of GHG-emissions reductions from such measures is often impossible, and it may be difficult to evaluate their cost-effectiveness
Non-state voluntary	Soft regulation; incentives; demonstration	New forms of responsibility and self-enforcement	Offer potential to reach sources of GHG emissions and vulnerability beyond the reach of municipal authorities. Scope to address climate change together with other issues of significance to non-state actors may provide incentives to pursue action and opportunities for addressing other challenges, such as social justice	Frequently small scale and with a basis on voluntary action, limited means to assess the contributions that are being made or for organizations to account for their actions. They may shift accountability from actors with official responsibilities for addressing climate change to those who have little in the way of power to address the issue
Public–private provision	Provision of low-carbon and resilient infrastructures, services and goods	Assembling and aligning diverse social and technical entities	Given the fragmented nature of urban infrastructures and service provision, this mode offers one means of creating opportunities for taking climate change into account in this landscape. Such partnerships may provide direct	Partnerships between public and private actors are far from a panacea, with concerns raised about the capacity requirements (in terms of co-ordination) and the fragility of such arrangements in the face of competing interests.

continued ...

Table 4.2 ... continued

Mode of governing	Policies and mechanisms	Logics and capacities required	Advantages	Limitations
			benefits, for example in terms of resources, knowledge and the pooling of different strengths	Partnerships can also be exclusive, and serve the interests of dominant groups within society while excluding the needs of the poorest or marginal. Partnerships also raise questions about the legitimacy and transparency of decision-making, and the extent to which decision-making is open and democratic
Non-state mobilization	Information and awareness raising; incentives; partnerships	Diverse modalities of power, including inducement, persuasion and seduction (Allen 2004)	Offer potential to reach sources of GHG emissions and vulnerability beyond the reach of municipal authorities. Can create broad-based political and social support for urban responses to climate change	As with enabling, mobilization efforts may be hampered by the challenges of engaging others in action on climate change. In addition, the mandate of non-state actors to call on others to act in relation to climate change and the extent to which they can be held accountable for doing so can be questioned. Furthermore, such efforts may serve to promote particular responses to climate change that accord with dominant social interests, perpetuating existing inequalities

Source: Adapted from Un-Habitat 2011: 108 and 112

its deployment is usually clear and transparent, and its intentions are explicit. This means that regulation can be very effective, especially in terms of targeting the use of particular technologies and encouraging behavioural change. On the other hand, its direct nature can also attract opposition, especially from those who might be adversely affected, and there are concerns that such measures are often regressive – that is, that they impose a disproportionate burden on the least well off in society. For example, when the mayor of London introduced a flat-rate congestion charge for driving in certain parts of the city, opponents suggested that this would place an unfair burden on the poorer sections of society.

The final mode of governing, *enabling*, reflects the changing nature of urban governance, set out in the Introduction to this chapter. In the context of limited direct powers in the urban arena, reduced capacity to provide infrastructure and services, and political opposition to forms of climate-change regulation, municipal authorities have turned to the enabling mode as one through which they can facilitate, co-ordinate and encourage the action of others through forms of partnership and engagement (Bulkeley and Kern 2006; Bulkeley *et al.* 2009; Gore *et al.* 2009; Hammer 2009; UN-Habitat 2011). Although this mode of governing has been particularly important in developed countries, evidence suggests that it is becoming an increasingly significant part of the ways in which municipalities and other actors seeking to govern climate change in cities in the Global South are responding to the issue (Bulkeley *et al.* 2009). Enabling can involve a variety of approaches, including information and education campaigns, incentives (e.g. subsidies, loans) and specific partnership schemes. This mode has frequently been deployed in the transport and building sectors, where behavioural change may have significant effects on the production of GHG emissions or levels of vulnerability. Seeking to develop and deploy an enabling mode of governing poses particular challenges for municipalities. Rather than relying on direct forms of power or intervention, such as is the case with self-governing and regulation, this mode requires the use of alternative forms of power, including 'inducement', such as financial incentives or the use of persuasion, and 'seduction', attempts to win hearts and minds (Allen 2004: 27–8), as well as 'generative power, the power to learn new practices and create new capacities' (Coafee and Healy 2003: 1982). In short, 'governments may have to cajole and persuade rather more than in the past, use skills of negotiation and diplomacy, involve more interests both inside and outside government, build coalitions of interests, and lead rather than direct' (Leach and Percy-Smith 2001: 4).

Whereas municipal modes of governing climate change were dominant during the 1990s, in keeping with the development of *strategic urbanism* as the emerging approach to urban climate-change responses documented above, new modes of governing initiated and pursued by private-sector actors are now visible (Bulkeley *et al.* 2009; UN-Habitat 2011). In these modes, non-state 'actors (such as foundations, development banks, NGOs and corporations) and public agencies outside the local authorities (donor agencies, international institutions) are initiating schemes and mechanisms for governing climate change in the city' (UN-Habitat 2011: 107) through three different approaches: *voluntary*, *public–private provision*, and *mobilization* (Table 4.2). In this context, a *voluntary* mode can be understood as encompassing the use of so-called 'soft' regulation to influence action within private organizations and public–private coalitions. The *public–private-provision* mode acknowledges the growing role of non-state actors in the delivery of infrastructures and services, either as a substitute for, or in parallel to, their provision by public agencies. *Mobilization* refers to efforts by non-state actors to encourage others to take action – for example, through the use of incentive schemes or education campaigns, mirroring the *enabling* mode discussed below (Box 4.3).

Drivers and barriers

The extent to which these different modes of governance have been deployed and have been successful is the result of a wide range of factors that can act as both 'drivers' and 'barriers' to achieving urban climate-change responses. For the most part, analysis has focused on the *institutional* and *political* factors that have shaped urban responses to climate change. However, a third set of factors, which can be termed *sociotechnical*, are also important (Table 4.3).

Broadly speaking, *institutional factors* can be regarded as those that shape the capacity of urban institutions – both formal organizations and more informal systems, codes and rules that guide social action – to respond to climate change. These factors include issues of knowledge, financial resources and the ways in which responsibilities for action are allocated and shared between different organizations. In terms of knowledge, research suggests that growing international scientific consensus about climate change has acted as a strong driver for urban responses, with cities adopting relatively stringent targets for GHG-emissions reductions. However, a lack of expertise has also proven to be a barrier, particularly in terms of adapting to climate change, where cities often lack the capacity and means to develop useful assessments of potential local impacts, as well as sufficient knowledge of the vulnerabilities facing their own populations to be able to develop local strategies. Challenges of data availability and access also exist in terms of

BOX 4.3

Manchester is My Planet: mobilizing the community?

In 2005, Manchester Knowledge Capital, a strategic partnership comprising universities, local authorities, public agencies and leading businesses in the Greater Manchester region launched Manchester is My Planet, supported by national-government funding. The initiative aimed to engage local communities and individuals in reducing their GHG emissions and asked people to 'pledge to play my part in reducing Greater Manchester's carbon emissions by 20 per cent before 2010 in order to help the UK meet its international commitment on climate change'. By the end of the campaign in 2009, organizers estimated that 21,309 local residents had taken the pledge, and that over 46,500 tonnes of CO_2 had been saved annually by residents taking action on their pledge commitments.

This case illustrates the potential for partnerships of public and private actors to mobilize members of the community to act on climate-change issues. However, there are a number of limitations to such schemes. First, measures to develop the knowledge and capacity of citizens to act have been limited, raising questions about the effectiveness of the pledge itself. Second, like all voluntary actions, undertaking the pledge carries no penalties for non-compliance. Third, the impact of such initiatives on reducing GHG emissions is difficult to determine. However, the scheme has been politically important. It has helped to establish the issue of climate change on local political agendas and beyond the confines of the municipal government, and it has been used as an example of best practice by national and other local governments, in turn serving to shape how initiatives have been developed elsewhere.

Sources: Jonathan Silver, Durham University, UK; UN-Habitat 2011: 114; Manchester is My Planet http://manchesterismyplanet.com/behavioural-change/the-pledge (accessed December 2011)

mitigation. Transnational municipal networks, such as CCP, Climate Alliance and C40, have sought to provide tools through which cities can assess their own GHG emissions, but this has been beset by problems of access to data that are either simply not available, particularly in less economically developed urban areas, or are owned by utility companies (Allman et al. 2004; Lebel et al. 2007; Sugiyama and Takeuchi 2008).

A further set of institutional factors relates to the multilevel-governance context within which urban responses to climate change are being developed. As set out above, this involves both 'vertical' relationships, between different levels of government, and various 'horizontal' interactions, between different partners and municipalities. Vertical forms of multilevel governance affect urban responses to climate change in three key ways: by framing the roles and responsibilities of local actors; by determining the duties and powers that municipalities have in key sectors (such as transport, planning, energy); and by enabling or constraining policy integration across these sectors (UN-Habitat 2011). These factors have been shown to be central to what it is that municipal governments feel able to do in response to climate change, as well as in shaping their ability to enact and enforce policy commitments (Betsill and Bulkeley 2007). Horizontal forms of multilevel governance, what are sometimes referred to as 'Type II' in the literature, have also been important in shaping urban capacity. These municipal networks – formal or informal associations between cities at a regional, national or transnational level and partnerships between municipalities and non-state actors – have been found to be important in developing the capacity of municipalities because 'they facilitate the exchange of information and experiences, provide access to expertise and external funding, and can provide political kudos to individuals and administrations seeking to promote climate action internally' (Bulkeley et al. 2009: 26; see also Granberg and Elander 2007; Holgate 2007).

Resources are also critical. In urban areas where municipalities lack the resources to provide even basic services, climate-change considerations are unlikely to gain much attention. Furthermore, the lack of such service provision can point to more fundamental issues, including the limited capacity municipalities have to invest in a context where 'local revenues go to recurrent expenditures or debt repayment', as well as a political attitudes that are 'unrepresentative, unaccountable and anti-poor – as they regard the population living in informal settlements and working within the informal economy as "the problem"' (Satterthwaite 2008a: 11). Where municipalities have economic resources, these may be organized in such a manner that makes expenditure on climate-related activities difficult, either because of the payback periods that are required or because of organizational structures that

Table 4.3 Drivers and barriers for urban climate-change responses

	Drivers	Barriers
Institutional	Availability of downscaled models of climate impacts for the city. Knowledge of social, economic and environmental vulnerability in the city. Capacity for monitoring GHG emissions. Pro-active national/regional government supporting local action. Co-ordination between different levels of government. Membership of transnational municipal networks. Formation of partnerships. Availability of external funding. Flexible internal-finance mechanisms	Limited capacity for creating or interpreting models for city/region. Lack of data or knowledge about vulnerability in informal settlements. Limited capacity for monitoring current/predicting future GHG emissions. Limited formal powers for municipal authority. Mismatches between responsibility for action and scale of problem. Disjointed work between sectors and absence of policy co-ordination. Lack of engagement by key business and civil-society actors. Limited financial resources
Political	Political champions. Recognition of co-benefits. Political will	Departure of key personnel. Political cycles/terms of office lead to short-termism. Prioritization of other policy agendas. Conflicts with other critical economic and social issues or sectors. Deliberate neglect of the marginal and vulnerable
Sociotechnical	Well-functioning and well maintained infrastructure networks. Emergence of niches/experiments for alternative technologies and social organization. New cultural and social perspectives on consumption/production in a post-fossil fuel era	Infrastructure deficit and failure to meet basic needs. Rigid infrastructure networks and institutional cultures that prevent change. Cultures of production and consumption based on continued availability of fossil fuels

do not allocate specific budgets to such cross-cutting issues. In some circumstances, municipalities have created novel financial mechanisms, such as revolving funds where the financial savings accrued from energy-efficiency measures are redeployed in additional energy projects, or forms of energy-service contracting where private companies invest in energy-efficiency measures and profit from the financial savings made (UN-Habitat 2011).

Political factors are those that refer to the possibilities for, and limitations to, responding to climate change that relate to the political leadership given to the issue and the political and economic context within which urban governance takes place, including issues of party politics and political terms of office, as well as more fundamental concerns, such as the role of specific economic and political interests in shaping pathways of urban development. Several studies have demonstrated that individual political champions or policy entrepreneurs (Box 4.4) have been critical to the development and pursuit of policies and projects at the urban level (Bulkeley and Betsill 2003; Qi *et al.* 2008; Schreurs 2008). Opportunities for municipalities and other actors to show leadership among peers is also important, such as the Forward Chicago initiative, established by the Climate Group and the mayor of Chicago and intended to engage Chicago's leading businesses in public–private partnerships to implement selected climate initiatives (The Climate Group 2011). The ability for political leaders to localize, reframe or 'issue bundle' climate change in relation to other local social and environmental benefits, such as reducing air pollution, improving neighbourhoods or saving money, has also been a critical factor shaping the political fate of climate change on urban agendas (Koehn 2008).

At the most fundamental level, political factors shape the debate about whether cities should or should not be addressing climate change. In many cities, the arguments 'not on my turf' and 'not in my term' are prevalent, particularly in developing countries where resources are limited and other concerns are more pressing (Bai 2007). Addressing climate change can often be in direct conflict with dominant urban political economies, where the very processes of increasing urban growth through land development, increased personal mobility and consumption that are regarded as necessary also serve to increase GHG emissions and leave issues of urban vulnerability to one side. These challenges may be particularly significant in less economically developed cities, where 'GHG mitigation has a negative connotation because of the perception that this will deny them of their basic right to growth in human services and economic activities; the prospects of "reduced growth" or "no growth" are not feasible' (Lasco *et al.* 2007: 84). Such tensions between dominant forms of urban growth and climate change are, however, also discernible in cities in developed countries. In both cases, it is arguable that the scope for municipal action on climate change is predetermined by existing ideas of what constitutes an appropriate form of growth, currently one that is dominated by a neo-liberal political and economic order (Rutland and Aylett 2008).

Sociotechnical factors refer to the combined effects of the material and technical conditions of cities – the means by which energy is produced, water is provided, buildings are constructed etc. – and the social, cultural, political and economic means that sustain and reproduce these urban systems. This combination of social and material factors co-produces the urban landscape within which urban responses to climate change take place, creating both possibilities for, and limitations to, how such responses are conceived and enacted. Urban responses to climate change therefore already take place within an existing urban morphology that is difficult to change – reducing GHG emissions from the transport sector creates a very different set of challenges in a dense European city than in a US city where urban sprawl is significant, or in rapidly growing cities in Asia where urban transport networks often do not keep pace with development. Equally, traditional practices of building design can provide significant barriers to the development and implementation of mitigation and adaptation measures, including what standards are used to determine levels of insulation, the amount of shading and indoor temperatures. In less economically developed urban areas, a chronic lack of infrastructure and shelter acts as a significant challenge for implementing adaptation measures. Furthermore, the infrastructure networks that are critical to achieving mitigation and adaptation goals – such as energy and water networks – often remain hidden from view, with decision-making power located outside the particular cities in which they operate. Such networks serve to embed particular forms of institutional and social practice – such as the ways

---

BOX 4.4

Policy entrepreneurs

Policy entrepreneurs, individuals involved in the innovation of policies or schemes, are defined by

> their willingness to invest their resources – time, energy, reputation, and sometimes money – in the hope of a future return . . . in the form of policies of which they approve, satisfaction from participation, or even personal aggrandizement in the form of job security or career promotion.
> (Kingdon 1984: 122)

in which we use water in our homes – and this combination of culturally adopted practices and large financial and political investments in particular forms of infrastructure provision can serve to limit the scope for change. However, it is possible to identify ways in which climate change is opening up and challenging such sociotechnical networks, through the development of, for example, alternative forms of energy supply, new cultures of consumption and alternative ways of fostering resilience.

Conclusions

Over the past two decades, the place of climate change on urban agendas has shifted dramatically. From being the concern of a few municipalities in Europe and North America, climate change has been championed by a range of cities in different political, economic, social and geographical contexts. This growth in the number of cities engaged in the issue has been accompanied by a shift in the nature of urban responses, from 'municipal voluntarism' towards 'strategic urbanism', as climate change becomes an issue that is of significance in relation to core urban concerns of growth and development. Municipalities and other urban actors now deploy an array of 'modes of governance' through which they are seeking to address both mitigation and, increasingly, adaptation. This is not to suggest that the development of urban responses to climate change has been uncontested. Research has identified a suite of challenges, from issues of institutional capacity to political economy and the sociotechnical networks through which urban responses are mediated, that have served both to drive and to constrain climate-change action.

These overall trends, of course, mask a significant diversity between and within cities in terms of how climate change is being addressed, and it is also important to remember that, for the vast majority of the world's cities, climate change is far from being a significant issue. For the most part, therefore, it could be argued that climate change remains simply 'un-governed' in cities – climate change is taking place, and processes of urbanization are a significant contributor to rising levels of GHG emissions (Chapter 3), but municipal authorities and other urban actors remain either unaware or unwilling to act. Likewise, as the effects of climate change begin to be felt in cities, incremental forms of coping and adapting are emerging, but, to date, concerted and comprehensive approaches to adapting to climate change at the urban scale remain few and far between. The next two chapters examine these challenges of mitigation and adaptation in more detail, and Chapter 7 considers the experiments and alternatives that may provide another means through which urban responses to climate change can be developed.

Discussion points

- Compare and contrast different theoretical perspectives on the concept of governance. How might urban responses to climate change be explained in each case? What are the differences and similarities between these perspectives? What are the implications for how we understand climate change as an urban problem?
- What are the core features that distinguish 'strategic urbanism' from 'municipal voluntarism'? What are the implications in terms of understanding the drivers of and barriers to urban climate-change responses?
- Taking one or more city as an example, consider the extent to which different modes of governing have been used by state and non-state actors in responding to climate change. What sorts of approach and action might be associated with the different modes of governing listed here? Why might particular modes be selected over others? What are the implications?

Further reading and resources

Examples of the initial urban responses to climate change and reviews of the history of urban climate-change responses can be found in:

Betsill, M. and Bulkeley, H. (2007) Looking back and thinking ahead: a decade of cities and climate change research, *Local Environment: The International Journal of Justice and Sustainability*, 12(5): 447–56.

Bulkeley, H. (2010) Cities and the governing of climate change, *Annual Review of Environment and Resources*, 35.

Collier, U. (1997) Local authorities and climate protection in the European Union: putting subsidiarity into practice? *Local Environment*, 2: 39–57.

Lambright, W. H., Chagnon, S. A. and Harvey, L. D. D. (1996) Urban reactions to the global warming issue: agenda setting in Toronto and Chicago, *Climatic Change*, 34: 463–78.

The core references listed in Chapter 1 provide detailed examples relating to the key contemporary challenges of responding to climate change at the urban level. For examples of how cities are responding to the issue, see:

- Asian Cities Climate Change Resilience Network: www.acccrn.org/what-we-do/city-initiatives
- Climate Alliance: www.klimabuendnis.org/member_activities.html
- Covenant of Mayors: www.eumayors.eu/index_en.html
- ICLEI CCP: www.iclei.org/index.php?id=800 (and links to regional campaigns).

5 Climate-change mitigation and low-carbon cities

Introduction

As global, national and local communities have sought to address climate change, the issue of mitigation – reducing GHG emissions or removing them from the atmosphere through the creation of 'sinks' – has been at the top of the agenda. As we saw in Chapter 1, the goal of the 1992 United Nations Framework Convention on Climate Change was to 'prevent dangerous anthropogenic interference with the climate system' through the reduction of atmospheric levels of GHG emissions (UNFCCC 1992). Subsequent international agreements, including the 1997 Kyoto Protocol and the 2009 Copenhagen Accord, as well as national-government targets, have also focused on this issue. GHG emissions have been monitored, predictions of future emissions have been calculated, and targets have been set. Despite these commitments, global levels of GHG emissions continue to rise. Even where national governments have signed up to such obligations, there is limited evidence that they have met their targets. Equally, where reductions have been achieved, this has often been as a result of factors beyond the direct scope of climate-change policy. For example, although the UK can claim to have cut emissions of GHG by 18.6 per cent over the period 1990–2008, these reductions were largely achieved through changes in the energy-supply system from coal- to gas-fired power stations, which were mandated because of other political and economic concerns (European Environment Agency 2011). In Germany, reductions of 22.2 per cent have been achieved over the same period, but this is partly a reflection of the changing boundaries of the country following the integration of West and East Germany in 1991 and the subsequent decline in economic growth and energy use in parts of former East Germany (European Environment Agency 2011).

It is against this backdrop that cities have started to address the issue of mitigation. The challenges experienced at the international and national levels have served to galvanize municipalities and a range of other urban actors to develop responses to climate change. In this sense, they have provided a very real driver for urban responses. At the same time, cities are not immune from the challenges that have been experienced at other levels of governance. In order to explore these issues, this chapter examines the responses of municipalities to the issue of mitigating climate change. Municipalities are an important part of the mitigation puzzle because they have jurisdiction over some of the key sectors where GHG emissions are produced – including urban development, the built environment, infrastructure systems (e.g. energy, water and waste), transport – as well as over green spaces, which may act as a 'sink' for carbon dioxide. The powers and capacities of municipal authorities in these areas vary considerably between and within countries (Bulkeley and Betsill 2003). Nonetheless, it is possible to identify some common approaches and measures that have been adopted across different cities. At the same time, it is important to remember that the geography of GHG emissions varies significantly between different urban places. As we saw in Chapter 3, to date, the vast majority of GHG emissions have been produced in cities in more economically developed countries, including North America, Europe and Oceania. Partly reflecting this geography, it has been in these regions that municipal authorities have historically been concerned with mitigation, although there is evidence that municipalities in developing countries are also beginning to engage with this agenda.

The rest of this chapter is divided into four main sections. The first considers what mitigation is, and the diverse ways in which this challenge has been tackled. The second examines in more detail the ways in which municipalities have sought to develop climate-change-mitigation policy. It outlines the approaches that have been taken and the types of action that have been pursued. The third section uses specific examples to illustrate the sorts of action that have been taken by municipal authorities and their partners in different sectors – urban development, the built environment and infrastructure systems. It considers how municipalities have sought to develop responses in these different fields, the benefits that they have achieved and their limitations. In the fourth section of the chapter, the drivers and barriers shaping the development and implementation of urban responses to climate-change mitigation are considered. The conclusions summarize the main issues considered in the chapter, as well as providing some questions of discussion, and further references and resources.

What is mitigation?

The main anthropogenic GHGs are carbon dioxide, methane, nitrous oxide and ozone, which arise from the burning of fossil fuels, agriculture processes and waste disposal, but other gases, such as chlorofluorocarbons and hydrofluorocarbons, that are synthetically produced for industrial processes and consumer products, including refrigeration, are also GHGs (see Chapter 1). In relation to climate change, mitigation refers to the processes through which the emissions of these GHGs are reduced, or existing GHGs in the atmosphere are removed.

The focus of mitigation activity, especially at the urban level, has been on the reduction of those GHG emissions that are produced from the burning of fossil fuels and, in particular, carbon dioxide. At its most straightforward, mitigation refers to reducing the overall amount of GHG entering the atmosphere. However, there are various interpretations as to what this means in practice. In some cases, absolute targets for GHG-emissions reduction are set – for example, that a municipality will reduce its 1990 levels of carbon dioxide by 20 per cent by 2020. In other cases, relative targets are set, where the goal is to reduce emissions in relation to predicted future levels, so that, for example, the target becomes one of reducing predicted emissions in 2020 by 20 per cent. Although this is still a reduction in the overall amount of GHG emissions that would enter the atmosphere, it usually entails an increase in GHG emissions from current or past emissions levels. Another way in which such relative GHG-emissions reduction targets are expressed is in terms of 'emissions intensity'. In creating GDP, particular consumer products, warming or cooling buildings, driving a car and so on, fossil fuels are used, and this results in GHG emissions. Mitigation efforts that focus on emissions intensity seek to reduce the amount of GHG emissions that are produced in the process of making particular products, providing thermal comfort, driving, or in creating economic productivity, so that, for any one unit, of, for example, GDP, the total amount of GHG produced is reduced. If GDP continues to grow (or we consume more products, drive further, build, heat and cool larger houses), the total amount of GHG emissions also increases but not by as much as if efforts to reduce the emissions intensity of these processes had not been undertaken.

A second means through which mitigation can take place is through the creation of 'sinks' – mechanisms with which GHG emissions can be captured from the atmosphere and stored. A common approach here, also adopted in municipalities, is the development of biological sinks – trees and other plant life that absorb carbon dioxide. Mitigation activities include afforestation, the deliberate creation of forested land, reforestation, the creation of forests on land that was previously forest but had been converted to other purposes, and deforestation, the deliberate prevention of deforestation activities for climate-protection purposes. Over the past 5 years, this last set of activities has become known as reducing emissions from deforestation and degradation (REDD) and is now an integral part of the international effort to address climate change through the UN Framework Convention on Climate Change. Alternatively, sinks can be created through carbon capture and storage, whereby carbon dioxide is captured at the point of production, such as at a large energy-generating plant, transported and then stored in a geological sink – underground rock formations that will enable the gas to be safely stored and not allow it to leak into the atmosphere. This set of technologies is currently at the early stage of research and development, and only a few trial sites have so far been tested.

Where mitigation efforts move beyond straightforward goals to reduce GHG emissions in absolute terms, and particularly where mechanisms to fund such initiatives are put in place, a whole host of challenges arise in determining just what counts as mitigation. The first arises in relation to assessing what is termed the additionality of initiatives – that is, whether they are efforts that are additional to what would have taken place in the absence of particular interventions or finance provided. A second, related issue concerns verifying whether measures have been put in place and whether they are achieving emissions reductions. A third set of issues relates to when, where and for whom setting relative emissions-reductions goals constitutes an acceptable response to the challenge of mitigation, given the complex issues of responsibilities for current atmospheric conditions and the challenges of addressing climate change and development goals simultaneously (Chapter 3).

These issues show that, although mitigation is at face value a relatively straightforward term, there is much at stake in terms of which GHG gases are included, if the focus is placed on avoiding emissions or on the creation of sinks, whether mitigation is approached in absolute or relative terms, and how additionality, verification and responsibilities are taken into account. Such issues are often not explicitly discussed in mitigation policy at the urban level, but underlying assumptions about what mitigation is, and is not, serve to frame how policy is developed and implemented.

Making mitigation policy

As Chapter 4 set out, historically, municipal climate-change policy has focused on mitigation. Over the past two decades, as a wider range of municipalities and other urban actors have come to engage with the climate-change agenda, a shift is discernible from a municipality-focused, voluntary approach, whereby municipal authorities have sought to reduce their own GHG emissions and enable others to take action, to a more strategic approach, in which climate change is regarded as central to urban development. This shift has been particularly prominent in large, global cities and in those municipalities that have pioneered climate-change responses, but, in other cities, the more municipal voluntarism persists. Despite the shift in the strategic importance of climate change, there remains continuity in terms of how climate-mitigation policy is framed and pursued. This section considers the approaches that have been developed for making climate-mitigation policy and the realities of implementing policy on the ground in the face of significant challenges.

Municipal-policy approaches: monitoring, targets ... action?

Initial responses to the climate change issue at the urban level began in the early 1990s (Chapter 4). Mirroring approaches that were being developed in the international arena by national governments, municipal authorities began to establish targets and timetables for reducing GHG emissions (Bulkeley and Betsill 2003). Although individual cities were responsible for pioneering this approach, what is striking about this period is the emergence of networks of municipal authorities operating transnationally to create a common response. While several different networks were important in creating this momentum, it was the CCP programme developed by ICLEI that established an approach to developing municipal climate-change policy that would serve as a model over the next two decades. From the outset, the CCP programme based its approach on a sequence of 'milestones' of monitoring, target setting, planning and implementation, an approach that was later adopted by another transnational, municipal climate-change network, the Climate Alliance (Box 5.1). Since these initial approaches were developed, the task of monitoring GHG emissions at the urban level has become both more complex and more contested (Chapter 3).

Despite these challenges, the basic approach developed over two decades ago – that municipalities should measure the GHG emissions that they produce and then seek to build policy responses on that basis – remains in place. Most recently, the C40 Cities Climate Leadership Group announced a new methodology for cities to account for their GHG emissions, developed initially in partnership with the CDP and now also with ICLEI. Underpinning this

> **BOX 5.1**
>
> **Milestone-based approaches to climate mitigation at the urban level: CCP methodology**
>
> Milestone 1: Conduct an emissions inventory for a baseline year (e.g. 2000) and a forecast year (e.g. 2020).
>
> Milestone 2: Adopt an emissions-reduction target for the forecast year.
>
> Milestone 3: Develop a local action plan . . . that describes the policies and measures that the local government will take to reduce GHG emissions and achieve its emissions-reduction target. . . . In addition to direct GHG reduction measures, most plans also incorporate public awareness and education efforts.
>
> Milestone 4: Implement policies and measures . . . [including] energy-efficiency improvements to municipal buildings and water-treatment facilities, streetlight retrofits, public-transit improvements, installation of renewable-power applications, and methane recovery from waste management.
>
> Milestone 5: Monitor and verify results as an ongoing process . . . providing important feedback that can be use to improve the measures over time.
>
> Source: Adapted from ICLEI 2012

approach over the past two decades has been a particular view of the policy process involved in responding to climate change, here captured succinctly by Mayor Bloomberg:

> 'I firmly believe that if you can't measure it, you can't manage it,' said C40 Chair, New York City Mayor Michael R. Bloomberg. 'That is true in business and it is true in government. Only by regularly and rigorously measuring and analysing our efforts can we learn what works, what doesn't and why, and take effective action.'
>
> (C40 2011a)

For Bloomberg, and other advocates of this approach, monitoring emissions provides the basis upon which municipalities should build their policy response. This is in keeping with other forms of public policy, in which indicators, targets, implementation and performance monitoring often provide the main elements of a policy cycle that can lead to successful delivery of strategies and objectives. There is evidence that such approaches are yielding results at the local level. A 2008 report from CCP Australia, for example, suggests that,

> in 2007/08 over 3000 greenhouse gas abatement actions were reported by 184 councils across Australia. Collectively these actions prevented 4.7 million tonnes of carbon dioxide equivalent (CO_2-e) from entering the atmosphere – the equivalent to taking over a million cars off the road for an entire year.
>
> (ICLEI Australia 2008)

At the international level, a 2006 report from ICLEI estimated that, across its 546 municipal members, the annual emissions reduced by CCP participants was 60 million tonnes of CO_2e, which amounts to a 3 per cent annual reduction among the participants and 0.6 per cent globally (ICLEI 2006).

Policy realities

Despite these upbeat accounts, taking such a policy approach to climate-change mitigation at the urban level has been challenging in several respects. First, despite the growth in the number of different models available at the urban level, assessing and monitoring urban-level GHG emissions remains fraught with difficulties, including how boundaries are assessed, which types of emission should be counted, and problems with data availability (Chapter 3). Second, unlike other policy areas where there is often a clear link between the institutional structure of the municipality and the issue at hand (for example, in education or health policy), climate change cuts across many established policy areas. This means that co-ordinating policy responses, in the way envisaged by a milestone-based policy approach, can be challenging. Some municipalities have established central units within their government structures for this very purpose; others prefer to seek to integrate climate-change activities into all relevant aspects of the municipality. Third, municipal authorities have highly varied levels of resource, capacity and responsibility for different sectors that contribute to GHG emissions within the city. For example, in some cities, municipalities may be the providers of energy and transport services; in others, these services are provided by the private sector. In other cases, national governments may mandate municipal authorities to undertake urban planning, whereas elsewhere this is a responsibility that falls to regional authorities. Operating across this diverse and uneven landscape has led cities to develop a range of policy instruments and measures through distinct modes of governing (Chapter 4), not all of which can be readily accommodated within this ideal of a policy process that moves from increasing levels of knowledge and target setting to implementation.

Although the monitoring, target-setting and performance-management approaches have provided significant impetus for municipalities to address their own emissions – through, for example, making the use of energy in municipal buildings visible, providing evidence of the financial case for energy savings, highlighting the benefits that could accrue from switching municipal fleets to alternative fuels – this has been harder to achieve beyond the confines of the municipality itself. As a result, analysts have found that, rather than following a smooth, linear path or policy cycle, municipalities have tended to adopt discrete policy measures on a case-by-case basis (Alber and Kern 2008; see also Chapter 7). Rather than serving as a strict set of steps, the policy approaches that have been developed in municipalities have served as a guideline and as a means through which their efforts in this area can be gathered together and represented to the outside world. Although the strategic urbanism approach that has developed in large urban areas and some pioneering municipalities has served to bring climate change closer to the heart of core urban-development concerns, there are few cases where this is leading to an integrated, long-term approach to developing new, low-carbon methods of urban development. Instead, iconic or emblematic projects are developed as a means of demonstrating the importance of climate change, and a range of other, smaller-scale projects and initiatives are assembled within current climate-change strategies in order to illustrate the potential of low-carbon urbanism (see the city case study on São Paulo).

CITY CASE STUDY

Climate-change mitigation in São Paulo

In 2009, São Paulo instituted its Climate Change Policy and, in doing so, became the first city in Brazil to take this step. The policy, also known as Law 14,933 of 5 June, explicitly recognizes the United Nations Framework Convention on Climate Change, while establishing the general objective of reducing São Paulo's GHG emissions. It includes an ambitious target of reducing GHG emissions by 30 per cent below 2005 levels by 2012.

São Paulo's Climate Change Policy was not the first local policy instrument to address climate change. The municipality had been experimenting with different mechanisms to respond to climate change since 2005, when it became one of the first Brazilian cities to prepare a GHG inventory. Energy use is responsible for most of São Paulo's GHG emissions, accounting for more than 75 per cent of the total. This is followed by solid-waste disposal, at about 23 per cent. These results mean that the GHG emissions resulting from land-use-change dynamics, agriculture and waste-water treatment are minimal. However, energy is a broad category, involving a wide and diverse range of key activities that underpin the functions of the city. The emissions inventory shows that the use of energy is linked to the city's electricity generation, industry, private and public transport and cargo and air transport, and to residential and commercial buildings. Although the large majority of the city's GHG emissions are related to road transport, the results of the inventory show the extent to which energy use, in its diverse forms, should be one of the city's main priorities in order to respond to climate change.

The 2009 Climate Change Policy was followed, in 2011, by a citywide Climate Change Action Plan, consolidating the issue of climate-change mitigation as an important local policy agenda. This emerging agenda has been useful, not only for highlighting the need to address environmental sustainability issues, but also for integrating other urban agendas that previously were treated in a disconnected manner. São Paulo's Climate Change Action Plan provides guidelines and priorities for future investment in the areas of transport, energy use, land use, building construction and waste-resource management. The guidelines, along with more than 100 specific actions accompanying them, range from the development of mechanisms to ensure that all public-transport vehicles run with clean, renewable technologies, to the development of experimental energy from waste initiatives. Other proposed actions are the implementation of voluntary energy-efficiency programmes for large consumers, the promotion of energy efficiency in public and private activities, the development of legislation to promote sustainable construction practices, and even the rearrangement of subregional urban centres to promote greater synergy between habitation and transport capacity.

Several of these actions are already under way. The municipality has installed biogas power plants at municipal landfills, improved energy efficiency in public buildings, installed street lights with LED technology, established bus rapid-transit (BRT) corridors and developed a network of bicycle lanes.

Reconfiguring its transport strategy is one of the main ways in which the city is responding to climate change. The municipality has a goal of replacing its entire fleet of buses with new vehicles running on renewable fuels by 2018. It is expanding the ability of the city's transport network to support electric buses and complementing the city's bus fleet with electric trolley buses, developed in partnership with private companies involved in the C40 Hybrid & Electric Bus Test Program. The replacement of the city's fossil-fuel-based bus fleet is not limited to electric vehicles; plans are under way for the use of a variety of biofuels, such as sugar-cane diesel, biodiesel blends and cane ethanol.

The municipality is also rapidly expanding its network of BRT corridors and aggressively promoting bus use: the authorities claim that investments in the city's transport are significantly increasing the amount of journeys made by bus, and, by 2020, they hope to see 70 per cent of all journeys made by bus. Other transport-related measures include the establishment of a comprehensive vehicle-inspection and maintenance programme, something that has resulted in a reduction of more than 41,000 tonnes of carbon monoxide and is likely to save over US$39 million in health costs by providing better air quality for the city's inhabitants.

The reconfiguring of the energy infrastructure is also being undertaken by promoting solar technologies, something that is being achieved via mandatory requirements for new buildings. Since 2009, all new residential dwellings with more than three bathrooms need to source at least 40 per cent of the energy

Figure 5.1 The City of São Paulo
Source: Andrés Luque

required for water-heating purposes through solar hot-water systems. Those homes with fewer than three bathrooms are required to make future provision for solar hot water by, for example, installing double piping (for cold and hot water). The mandatory requirements for solar hot water in new construction also cover hotels, health facilities, sport clubs, schools and nurseries, among other types of building.

São Paulo's Climate Change Action Plan was prepared in partnership, involving a multiplicity of government agencies as well as representatives of civil society, academia and the private sector. It was developed by the city's Climate Change Committee through its six working groups: energy, construction, waste, public health, transport and land use. The city's Department of Urban Development played a key role by co-ordinating the activities of the committee, but other government departments, as well as external organizations, contributed and played a variety of roles, including the city's Departments of Finance, Transport and Environment, the Department of Energy of the State of São Paulo, the National Association of Vehicle Manufacturers, the Federation of Industries of the State of São Paulo, the Bill Clinton Foundation and several energy utility providers. This case demonstrates

Figure 5.2 Solar hot-water systems in São Paulo's social-housing projects
Source: Andrés Luque

that, even where municipalities are leading the formation of climate-change plans, their actions are often the result of the formation of a coalition of actors who support and implement mitigation measures.

At the global level, São Paulo has also played an important climate-change leadership role. In 2011, São Paulo hosted the C40 Mayors' Summit, the biennial meeting of the Large Cities Climate Leadership Group. The meeting was hosted by the city's mayor, Gilberto Kassab, and counted with the participation of the mayors of Sydney, Addis Ababa and New York, alongside public representatives of nearly thirty-five other cities. The summit finished with a joint statement where participating municipalities highlighted the role that megacities play in climate change, the need for these cities to have a voice in the debate, the important role of national governments and international treaties in empowering the leadership of these cities, and the need for municipalities to have sufficient support and resources to take action.

Andrés Luque, Department of Geography, Durham University, UK

Focusing on a policy approach where significant emphasis is placed on monitoring existing GHG emissions on the one hand, but where climate-change strategies and measures are based on the accumulation of existing projects and initiatives where municipalities can have a direct influence, on the other hand, risks masking three critical issues. First, the number of measures being implemented by municipalities – although growing – remains low. A pilot project undertaken by the Carbon Disclosure Project with eighteen CCP cities in the USA found that, although commitment to targets is high, the number of measures that have emerged as a result of the policy approach is currently very low (Carbon Disclosure Project 2008). Second, despite its weaknesses in terms of implementation, such a policy approach has been a powerful device for framing what it is that municipal authorities have to be concerned about with respect to mitigating climate change. By focusing on those emissions that are produced within the city or through its use of energy, Scope I (see Chapter 3), this framing excludes the emissions implicated in the consumption of goods and services within the city. In the process, this may mean that certain policy areas where municipalities have some jurisdiction or influence are being neglected. It also means that some of the most significant means through which cities contribute to climate change – through the consumption of goods and services, international trade and transportation, for example – are excluded from the agenda (Rutland and Aylett 2008). Third, despite renewed strategic concern for climate change, and the prominence it is afforded on the urban agenda of some globally important cities, such strategies remain rather incremental in their approach and disjointed from areas of urban policy development.

Developing a viable and robust approach to climate-change mitigation at the urban scale is therefore far from straightforward. However, as the number of cities engaged in climate-change mitigation has grown, and the extent of their commitments to action has increased, a wealth of evidence has been accumulated about the range of measures and initiatives that are being developed. Although these may fall short of the ideal models of the policy process discussed above, they do suggest that mitigation has become a very real issue in many cities across the world.

Municipal climate-change mitigation in action

> The kind of climate change initiatives that local governments can most easily do appear to be such activities as climate change and renewable energy target setting, energy efficiency incentive programs, educational efforts, green local government procurement standards, public transportation policies, public–private partnership agreements with local businesses, and tree planting.
>
> (Schreurs 2008: 353)

As this analysis indicates, the ways in which municipalities might seek to develop and implement the sorts of policy approach articulated above are potentially never-ending. From planting trees to changing light bulbs, from retrofitting buildings to developing new infrastructure systems, the scale, nature and complexity of initiatives that have been undertaken vary enormously. This is in part a reflection of what it might be possible for municipalities to do in different countries. Municipal competencies – their powers and responsibilities in any particular domain – vary considerably between different sectors and countries. Particularly in cities where these competencies are limited, municipalities have focused on the measures that can be undertaken in a self-governing mode, where municipalities seek to address their own production of GHG emissions in buildings and vehicle fleets and in the (usually limited) infrastructure networks and services that they provide (Chapter 4). This is not an approach that has been confined to cities in more developed economies: for example, in Cape Town, a target of increasing energy efficiency within the municipality by 12 per cent by 2010 was set, and, in Yogyakarta, a programme to retrofit lights and reduce air conditioning in government buildings was established in 2003. As Chapter 4 set out, municipalities also deploy other modes of governance in order to undertake emissions-reduction activities beyond their own operations. A survey by Arup of the C40 cities network (Arup 2011a) found that municipal competencies to address GHG emissions were strongest in the transport, waste, water and land-use-planning sectors, where municipalities either owned and operated infrastructure networks or were able to set strong regulations. In the building sector, several municipalities also had strong powers to regulate the private sector and/or owned municipal housing. Municipal powers were weakest in the energy sector, where municipalities reported powers to shape the vision of energy-supply systems but had little direct control or influence. In other cities, where resources and power are less concentrated, research has found that municipalities have more limited direct powers to regulate and provide services and, instead, tend to rely on an enabling mode of governing in order to address climate change (Bulkeley and Kern 2006).

Municipal competencies and the modes of governing that they deploy vary between cities and across the three main sectors – urban development, the built environment and infrastructure systems (energy, water, waste and mobility) – where actions have been forthcoming. These actions range from those that seek to operate at a whole-city scale, within particular neighbourhoods, or those that target individual households or businesses. Across these different sectors, municipalities have also demonstrated various levels of commitment to acting on climate change – from those actions that show leadership on the issue, to those (much less common) actions that involve undertaking deep cuts in GHG emissions.

Managing urban development

In different regions of the world, cities are grappling with complex urban-development challenges, including rising levels of urban sprawl and the growth of informal settlements. In this sector, the mitigation challenge relates primarily to those cities experiencing the growth of formal housing and commercial development, predominantly at the urban fringe. How and where new urban development takes place has implications for the amount of energy-intensive materials used to create new urban landscapes (e.g. concrete, steel), the reliance on private motorized transport (which is currently dominated by the use of fossil fuels), as well as the energy that will be used in such buildings over their lifecycle. In Chiang Mai, Thailand, for example, research found that 'the ribbon and spike sprawl pattern of urban and commercial development . . . together with growing economic prosperity, has

... created a surge in personal vehicle use for going to and from work and markets' (Lebel *et al.* 2007: 101). In other cities, the primary pressure on urban development comes from the growth of informal settlements, sometimes referred to as slums. Recent estimates suggest that the global number of slum dwellers will increase to 2 billion before 2030 (UN-Habitat 2008). Although the GHG emissions associated with such forms of urban development are negligible (Chapter 3), such areas are particularly vulnerable to the effects of climate change (Chapter 2) and also experience limited access to affordable and adequate shelter and energy services. Addressing these challenges of urban development through the perspective of climate change has the potential, not only to reduce GHG emissions in areas of high levels of urban growth and energy consumption, but also to create more resilient forms of development that are also able to deliver the services that people need in a low-carbon manner.

Municipal competencies can be significant in relation to urban development. Those municipalities with planning responsibilities can establish land-use or zoning plans that set out how different areas of the city and its fringes should be used. These planning powers usually operate across the whole city, with municipalities able to mandate the sorts of development that take place – for example, requiring different levels of housing density, parking-space allocations or mixed-use developments – and to specify some aspects of building design. In the UK, for example, national planning guidance requires local planning authorities to take climate change into consideration in decision-making (Bulkeley 2009). In other cities, including Melbourne, São Paulo and Cape Town, municipalities have adopted principles of compact city planning, which advocate high-density development and mixed-land-use principles, so that residential and commercial parts of the city are proximate, and commuting distances can be reduced. However, such planning powers are not universally enjoyed by municipal authorities, and, in some cases, planning decisions are taken at regional or even national scales. Even where planning powers do exist, there can be significant challenges associated with their implementation and enforcement. In the US, for example, urban sprawl can take place at the boundaries between municipal jurisdictions. In other cases, the problem lies with enforcement. In Jakarta, Sari (2007: 141) found that,

> while [the] zoning permit is theoretically supposed to be a tool to control land use, in reality corrupt practices have rendered it ineffective. A 1993 study under the Jabotabek Management Development Project shows that there are many developers that are not in compliance with the existing land use allocation.
>
> (Sari 2007: 141)

Furthermore, whether models that advocate compact city development are appropriate in all urban contexts is debatable, especially where density levels are often already high, leading to overcrowding, and urban conditions are inadequate to meet basic needs.

Alongside these efforts to plan and manage existing urban growth, specific low-carbon urban-development projects have also emerged. These include projects to develop carbon sinks within cities through tree-planting schemes, to reclaim and reuse brownfield land within cities, as well as the building of new urban districts or entirely new, low-carbon cities. One example is the Dockside Green development in Victoria, Canada, which has been built around the principle of providing economic, environmental and social value, and on the basis of a 'total environmental system' that creates a 'self-sufficient, sustainable community where waste from one area will provide fuel for another' (Dockside Green 2012). Sustainable-development principles have also been incorporated in the plans for the redevelopment of the site of Hong Kong's Kai Tak airport, which will include green space, energy-efficient design, public transport and a district cooling system that will reduce the amount of energy used in air conditioning. Perhaps one of the most iconic examples of this form of low-carbon urban planning is that of Dongtan, an eco-city development on the outskirts of Shanghai that aims to be self-sufficient in energy and water, with local food provided from surrounding agricultural land (Hodson and Marvin 2010: 69). Despite, or perhaps because of, its visionary approach, the project has met with significant delays, and, by 2009, such little progress had been made that some have suggested that it will never be implemented (Pearce 2009). Furthermore, developing greenfield, urban fringe sites as a means of mitigating climate change can also be questioned, given that they have the potential to generate significant GHG emissions from the materials used in new construction and in terms of the mobility requirements that are generated by such locations, and that they can serve to exacerbate social inequalities because they are often exclusive in nature (Bulkeley and Castán Broto 2012b). Although such developments may offer an approach that can bridge the gap between the needs for urban growth and expansion and climate change, they may also serve to exclude other possibilities for urban development that seek to reuse existing areas of the city or seek to encourage the urban growth of other cities.

Beyond urban planning, some cities have adopted broader-scale principles and approaches in order to develop a low-carbon approach to development. In China, a specific programme for low-carbon cities was launched by the national government and WWF in 2008, with the aim of developing a low-carbon economy and low-carbon lifestyles among residents in Shanghai and

Baoding (Liu and Deng 2011; Hodson and Marvin 2010: 78–9). Similar principles have been adopted by a range of cities. In Vajxo, for example, the municipal government, along with other public and private partners in the city, has developed a plan to become a 'fossil-fuel-free' city, through the development of alternative sources of energy, compact urban planning and incentives for behavioural changes that reduce energy use (Gustavsson et al. 2009). The London Development Agency advances an explicitly economic approach to taking such measures, suggesting that, as a market leader in some sectors of the low-carbon economy, London is in a good position to take advantage of new flows of investment in this economic area. In order to capitalize on these opportunities, the LDA is 'helping to stimulate and drive growth in the low carbon and environmental goods and services markets, creating sustainable jobs and opportunities for Londoners', and has established a Green Enterprise District in the Thames Gateway, 'where sustainable industries will exist alongside sustainable communities and infrastructure' (London Development Agency 2012). In Cape Town, the opportunities presented by a low-carbon economy are also regarded as significant: 'moving our city towards a low carbon economy, using renewable and green technologies, provides massive opportunities for job creation, skills development and poverty alleviation' (City of Cape Town Environmental Resources Management Department 2009). Such approaches illustrate how climate change has become a more strategic issue, whereby it is not regarded only as one issue among many that need to be addressed, but rather becomes central to how urban development should take place (Chapter 4).

Designing and using the built environment

While urban planning and development provide the means of tackling the challenges of urban growth and achieving low-carbon economies, addressing mitigation across the city also requires specific engagement in particular sectors. The design and use of the built environment, including public (e.g. government offices, hospitals), domestic (housing) and commercial (e.g. offices, factories) buildings, are critical areas for mitigation. The use of energy within the built environment is the result of complex interactions among building materials, design, the systems used to provide buildings with energy and water, and the ways in which buildings are used on a daily basis (Foresight 2008). Globally, this sector 'consumes roughly one-third of the final energy used in most countries, and it absorbs an even more significant share of electricity' (Bulkeley et al. 2009: 43). It has been in this area, and in particular on the issue of energy efficiency, that most municipal effort has been concentrated (Bulkeley and Kern 2006). A focus on energy efficiency has been found to be capable of advancing 'diverse (and often divergent) goals in tandem' (Rutland and Aylett 2008: 636), serving to align climate-change policies with more long-standing concerns, including, for example, the efficient use of resources, addressing problems of cold homes, and financial savings. Within this arena, municipalities have sought both to affect how the built environment is designed, through the use of standards, regulations and the retrofitting of existing buildings, and to shape how energy is used within buildings, through economic incentives and by providing information to households and businesses about how they might change their behaviour.

Traditionally, municipal authorities have had limited powers to affect the design of the built environment at the city scale, although they frequently have the authority to approve or disallow applications for alterations to specific buildings. However, there is evidence that such standards and regulations are now being put into place. One example is the use of solar thermal ordinances in cities such as Thane, India, São Paulo, Brazil, and Barcelona, Spain, which mandate the use of solar hot water in new buildings. In other cities, specific standards for energy efficiency are now being mandated, for both domestic and commercial buildings. For example, in 2005, under the Planning and Environment Act 1987 (Victoria), the City of Melbourne introduced the C60 Planning amendment to stipulate that all new office developments over 2,500 m^2 must meet an energy star performance requirement of 4.5. Although such standards frequently apply only to new buildings, they can also be used where buildings are being substantially refurbished. In addition to using regulatory powers, municipalities have also sought to show leadership in this area by building or retrofitting buildings as demonstration projects that show how particular energy-efficiency and low-carbon technologies can be used in practice. One such example is the development of solar-powered air conditioning for a hospital in Thane (Box 5.2).

At the neighbourhood scale, municipalities are often involved in efforts to 'retrofit' existing buildings in order to improve their standards of energy efficiency. Research has found that such efforts tend to be focused on municipally owned, residential buildings, with initiatives undertaken in various European cities, including Vienna (Austria), Stockholm (Sweden), London (UK), Munich (Germany) and Rotterdam (the Netherlands), as well as in the United States, where funding for 'weatherization' projects is provided by the national government, for example in New York, Chicago and Philadelphia (UN-Habitat 2011: 96). Such programmes and initiatives are often regarded as having 'win–win' potential for addressing climate change, because, they not only reduce the amount of energy needed for heating and

BOX 5.2

Thane's solar hospital

The city of Thane, in Mumbai's Metropolitan Region, has been committed to renewable energy and energy efficiency since the early 2000s. Thane, with 2 million habitants and continued growth owing to its proximity to Mumbai, is taking active steps to reduce India's energy gap between supply and demand. Among other actions, the city has implemented energy audits in key municipal services and offices, installed solar-powered traffic lights and photovoltaic systems for the main municipal offices, and enacted a policy making the use of solar hot water mandatory for all new construction. Although energy conservation is the city's main motivation, the city has also prepared a citywide energy audit accounting for its carbon emissions.

In practice, Thane is developing models for low-carbon innovation at the city level. Between 2010 and 2011, the municipal government installed a solar air-conditioning system at the city's main hospital, replacing an outdated and inefficient traditional air-conditioning system. This system works with over ninety Scheffler parabolic reflectors located on the hospital's roof. These parabolas generate steam, which is then directed to a vapour-absorption unit (heat exchanger) for the generation of cold air. As the hospital only provides air conditioning to the operating rooms and sensitive medical areas, the sun provides the required energy for most of the day. For those moments when the sun is not shining fully (such as mornings, late afternoons and cloudy monsoon days), the system operates with a boiler running on 'agro-briquettes', a biological fuel made from agricultural waste.

However, no innovation comes without the risk of failure. Thane's municipal government managed to distribute the risk associated with this innovation by partnering with the regional and national government bodies for funding, and linking the system's performance to the schedule of payments of the private contractors delivering the system. The latter, in turn, had a keen interest in achieving success in order to showcase one of the first solar air-conditioning systems in the world and, in this way, expand business opportunities.

Andrés Luque, Department of Geography,
Durham University, UK

cooling buildings, but they also address issues of energy affordability, ill health and broader issues of social and economic well-being through providing decent housing. However, where the goals of such interventions are explicitly focused on improving the levels of energy service that 'fuel-poor' communities and households receive, the potential for achieving an overall reduction in energy consumed (and, therefore, of GHG emissions) may be reduced, because of the priority for warmer or cooler levels of indoor comfort. Achieving both improved energy services and a reduction in overall energy use to achieve climate-change-mitigation benefits may therefore require rather ambitious programmes. Because of the considerable challenges of energy affordability and access that such cities face, research suggests that such retrofitting programmes are not common in cities in the Global South. Where action is taking place to address energy use in the built environment, 'the use of energy efficient materials has been an important means through which municipal governments and other actors have sought to address GHG emissions reductions and the provision of low-cost housing to low income groups', for example in Buenos Aires, Argentina, and Rio de Janeiro, Brazil (UN-Habitat 2011: 97).

It has, perhaps, been in the built-environment sector that municipalities have taken the most significant steps towards changing the ways in which energy is used. Often termed demand management, efforts to improve energy efficiency and reduce the overall use of energy depend not only on the physical attributes of buildings and appliances, but also on how they are used – however efficient a building might be, inhabitants turning the heating system on and wearing a t-shirt in winter will still lead to an 'inefficient' level of energy use. Municipalities have used two main approaches in seeking to manage demand – economic instruments (e.g. taxes, fines, incentives) and information campaigns of various kinds. Tokyo municipal government is unique in having developed a city-level emissions-trading scheme that sets caps on the level of GHG emissions that large commercial operations and factories can produce and allows them to trade with one another in order to achieve the necessary reductions in the most cost-effective manner (see Chapter 7). More commonly, municipalities have sought to target individual households and commercial operations through providing them with incentives and information for reducing energy use. For example, in Hong Kong, the government suggests that thermostats on air-conditioning systems should be set to 25.5° in order to conserve energy, while in Newcastle, Australia, residents are able to borrow an Energy Reduction Kit that provides them with a means of monitoring energy consumption and indicating areas where changes might be made. Although such interventions are extremely well intentioned and may provide a means by which individuals can reflect on their energy use, they subscribe to a model of how and why people might change their energy behaviour that some authors have termed the 'value–action' gap (Blake 1999; Hargreaves *et al.* 2010). In this approach, cognition of, or awareness about, an issue is seen to be linked to behaviour or action. Although people express concerns about climate change, opinion polls and rising trends of energy use suggest that they often fail to act upon these. This gap between values and action is then seen to be a matter of enhancing awareness – increasing people's knowledge about an issue will lead to action. However, this approach overlooks the ways in which energy use is tied up with a set of social practices that are underpinned by a range of other values and beliefs and that are structured by the ways in which energy services are provided and consumed (including networks, artefacts, buildings, social norms and so on) (Shove 2003, 2010; Hargreaves *et al.* 2010). This alternative perspective, sometimes referred to as a social-practices model, suggests that those municipal authorities seeking to reduce demand at the level of individual households and commercial operations need to engage with how such energy services are produced and consumed, to understand what factors, aside from information, are required in order to develop low-carbon forms of energy practice.

Reconfiguring urban infrastructure networks

Urban infrastructure networks are important, though often neglected, components of urban GHG emissions. The type and carbon intensity of energy, water and waste services used within cities shape the extent of GHG emissions produced. Such networks often lie outside the direct control of municipal authorities, can be politically contentious, and often require long-term planning and investment. At the same time, the emergence of a splintered landscape of infrastructure provision in developed countries, together with the enduring fragmentation of such networks in cities in the Global South, is creating new opportunities and challenges (Graham and Marvin 2001). Within this context, mitigating climate change is becoming an important issue, but one that competes for attention with other pressures for the security and affordability of infrastructure networks and the provision of basic services. In particular, addressing climate change has come to be a key issue in the transport sector and to form a critical part of decisions over current and future energy provision for cities, alongside new concerns for the continued security of energy supply in the face of rising energy prices and concerns for the future sustainability of fossil-fuel-based energy sources. To date, there has been less concern, at the municipal level, with the energy-intensity of water, waste and sanitation systems, although rising energy costs and concerns about future water shortages, coupled with new approaches that view waste as an (energy) resource, are beginning to surface.

Urban mobility is supported by a wide range of infrastructure networks, including the physical structures that support trains, cars, buses, cycling and walking, as well as the regulations, incentives, institutions, cultural norms and physical things (trains, cars, bikes and even shoes) that serve to structure transport choices. Existing patterns of urban development and morphology, whether cities are compact or sprawling, together with the dominant forms of infrastructure provision, shape the extent to which urban mobility contributes to GHG emissions. Globally, the transport sector contributes roughly a quarter of all GHG emissions, and, although this is currently higher among OECD countries, it is rising in the developing economies, particularly India, China and Brazil, as a result of growing urban sprawl and the modal shift away from cycling and walking to motorized vehicles that is taking place rapidly in these countries (Short *et al.* 2008). Beyond the sorts of urban planning and development initiatives discussed above, municipalities have sought to provide public transit systems, to promote low-carbon vehicle technologies and to encourage individuals to choose alternatives to the private vehicle. In terms of public transit, one approach that is growing in popularity is the BRT system, which involves guided bus ways or dedicated bus lanes, together with mechanisms to give buses priority in urban traffic. In their survey of C40 cities, Arup (2011: 32) found that such systems have been introduced in thirteen cities, with a further eight planning their introduction, with all six C40 members in Latin America having introduced or planning such a system. The emergence of such systems in Latin America can be traced to the pioneering experience of Curitiba and Bogotá and the process of knowledge sharing that has taken place through the C40 and other networks, as well as the provision of international financial support for the development of BRT systems within Latin American cities. Such initiatives often depend upon a partnership of municipalities and the (often privately owned) providers of public-transport services, as well as access to sufficient capital to establish what are often costly schemes.

By contrast, efforts to promote alternative vehicles and to shape the demand for transport can involve relatively low upfront costs. In numerous cities, efforts to promote hybrid, hydrogen and electric vehicles are taking place. Although, initially, such schemes focused on the development of new technologies within municipal fleets, such as schemes to use biofuels for the public-transport fleet in San Francisco (Hodson and Marvin 2010: 73) or fuel-cell buses in Berlin and Hamburg (UN-Habitat 2011: 102), more recent initiatives have focused on enabling individuals to use alternative vehicles. For example, Source London was launched in 2011 as an electric-vehicle charging-point network for members with electric cars to develop the number of charging points in the city and improve access for electric-car owners (Source London 2012). In Paris, the Autolib scheme launched in December 2011 is a 'pay-as-you-drive' scheme for electric vehicles that follows a model developed under the Zen Car scheme in Brussels, whereby individuals subscribe to a car club and are able to book cars located at a number of points around the city. Alongside efforts to promote these alternative vehicles, municipalities have also sought to promote the use of other modes of transport, particularly cycling, through the improvement of cycle networks and the development of cycle-hire schemes. Such schemes are particularly popular in European cities, including the Bicing initiative in Barcelona, the Romainbici scheme in Rome, and Paris's Vélib. In addition to seeking to offer alternative transportation choices, some cities have sought to manage or restrict demand for the use of private vehicles. Such measures are usually highly politically contentious and difficult to implement. Nonetheless, a survey of C40 cities found that twenty-three cities had implemented one or more forms of transport-demand management, which included congestion charging, restrictions on the use of vehicles by time/day, or parking restrictions (Arup 2011: 28). Although such schemes are often developed in order to address concerns about congestion and air pollution, they can also serve to encourage shifts to lower-carbon modes of transport. However, where such restrictions are based on congestion charges, they may serve to exclude those who cannot afford them, while granting others a licence to pay and pollute.

Energy systems, those infrastructures that generate and supply the energy used in buildings and commercial and industrial enterprises, are the source of the majority of urban GHG emissions. Although municipalities often exercise little control over the ways in which energy is generated, distributed and consumed within cities, two main approaches for developing low-carbon forms of urban energy network can be identified. First, many municipalities have sought to reduce the carbon intensity of their existing energy infrastructure. One set of infrastructures that is the direct responsibility of municipal authorities in many parts of the world is street lighting. Lighting may account for up to 10 per cent of global GHG emissions, with street lighting accounting for 8 per cent of this total (Hoffmann 2011: 84). Addressing the energy used in street lighting may therefore be one way in which municipal authorities can cumulatively make an impact in this area. Over the period 2001–6, Yogyakarta developed a Streetlight Management Scheme through the CCP programme, which 'involved the retrofit of 775 light bulbs and the installation of 400 energy meters at a cost of $1.7 million, resulting in an annual saving of 2,051–3,170 tonnes of carbon dioxide (annual energy saving 4,278,408 kWh) and an estimated $211,765' (Bulkeley et al. 2009: 68). Similar schemes have also been developed in Mumbai, India, Beijing, China, and Melbourne, Australia.

In other parts of the world, cities are also responsible for urban-based combined heat and power (CHP) and district heating schemes, which provide different combinations of heating (and sometimes cooling), hot water and electricity to different parts of the city. Berlin, Germany, for example, has an urban heat network that includes over 1,500 km of pipes and over 280 district-scale CHP plants. In August 2008, municipal authorities, together with other private-sector partners, launched the CHP Pilot City Berlin, with the aim of expanding the urban CHP network from 42 per cent of the heat market in 2008 to 60 per cent in 2020 (Neuhäuser 2010). In London, the LDA has initiated the Decentralised Energy and Energy Masterplanning (DEMaP) programme in order to show where existing heat networks are and 'to assist both public and private sector to identify Decentralised Energy (DE) opportunities in London', which in turn will 'contribute towards the Mayor's target of providing 25 per cent of London's energy supply from decentralised energy sources by 2025' (Greater London Authority 2012).

A second approach that municipalities have developed is the creation of new forms of low-carbon energy systems. Particularly in cities in the Global South, climate-change mitigation is expressed as an additional benefit alongside primary objectives of increasing energy security, for example in Quito, Ecuador, Bogotá, Colombia, and Rio de Janeiro, Brazil, where initiatives seek to reduce dependence on oil by promoting the use of natural gas in households (UN-Habitat 2011: 99). Energy-from-waste plants are also emerging as a response in such cities to the triple challenge of affordable energy services, energy security and growing post-consumer waste. An important driver here has been the CDM (see Chapter 1 and Chapter 7). In other cities, the focus has been on the development of new renewable technologies. In the US, for example, the Department of Energy's Solar American Cities programme seeks to develop solar-energy-demonstration projects in partnership with local authorities (Chapter 7). Alongside such projects, which focus on the development of one renewable energy technology, are 'smart city' projects, which seek to develop electricity-grid systems that enable demand for electricity to be more closely matched with available supply, through the use of smart meters, incentives for using electricity at particular times, and new storage technologies. By seeking to match demand with supply, smart grids are designed to be better able to address the intermittent nature of many renewable-energy sources and to use them more efficiently (Box 5.3). Such projects are still in the early stages of development and have yet to be tested or assessed. Given that they require rather different forms of behaviour by consumers and energy providers, it is likely that, in practice, realizing their benefits will be more challenging than is assumed by the rhetoric of existing demonstration projects.

Although different in scale and scope, efforts to reduce the carbon intensity of existing infrastructure and to develop new energy systems for cities are characterized by concerns both to secure energy supplies and to leverage economic benefits through carbon control (Hodson and Marvin 2010; While *et al.* 2010). For some municipalities, these drivers have been manifest in explicit calls for energy independence. In San Francisco, an energy-independence ordinance required municipal departments to prepare 'an Implementation Plan to solicit 360 MW' of new generation and energy savings that would contribute approximately a third of the city's electricity needs (Hodson and Marvin 2010: 85). Spurred by national legislation, other cities in the US have followed suit in setting out strategies through which to reduce dependence on fossil fuels. Although much less extensive, similar approaches that seek to develop water-conservation measures and to tap into previously unused forms of urban water, through recycling water that has previously been used ('grey' water) or capturing rainwater, have also been used as a means of reducing the GHG emissions involved in the production of clean water and in promoting urban self-sufficiency (see also Chapter 6). In Hammarby Sjöstad, Stockholm, the redevelopment of a brownfield site has been undertaken based on principles of closed-loop recycling, so that all wastes are converted into useful resources within the development, and rainwater and storm-water are used for heating, cooling and power generation (Coutard and Rutherford 2010). Although the Hammarby Sjöstad case is a particularly famous example, similar principles of promoting self-sufficiency in water use can be found elsewhere. In Sydney, the Olympic Park has become the site for the development of a water-reclamation and -management scheme that 'recycles water from sewage and storm-water to supply irrigation, ornamental fountain and toilet flushing applications across Sydney Olympic Park and in the suburb of Newington' (Sydney Olympic Park Authority 2012; see also Hodson and Marvin 2010: 75). Such approaches, which aim to promote energy and water security, contribute to what Coutard and Rutherford (2010) term 'post-networked urbanism', where different forms of infrastructure network and service provision are emerging as an alternative to large-scale integrated grids. Although these moves do not reflect a wholesale shift to decentralized technologies and are often highly contested, they do demonstrate that the drive to mitigate climate change is leading to new forms of urban development.

BOX 5.3

Smart cities: low-carbon future?

The term 'smart cities' is usually used to describe urban initiatives where information systems (often information communication technologies (ICT)) support city-management processes. Smart-city projects include a wide diversity of projects, such as new urban developments where ICT plays a key role, e-governance projects, city way-finding and signage initiatives, electric mobility and smart-energy projects.

The cities of Boulder (Colorado, US) and Amsterdam (the Netherlands) are experimenting with pioneering approaches to become 'smart cities'. Boulder's Xcel Energy, the local energy provider, has championed the SmartGridCity project. This is a smart-grid pilot rollout, enabling two-way, high-speed digital communication between all the components of the city's energy system, including the users. Whereas traditional energy meters are read once a month, Xcel's smart meters are read every 15 minutes. This means that homes and offices are sharing information about their energy needs and use in real time, allowing the utility to predict outages, monitor the health of the system and adjust generation and consumption to increase overall efficiency. As with any increase in efficiency, the resulting carbon savings can be significant. The smart grid also enables greater integration between decentralized renewables and the city's energy system, something that is seen by Xcel Energy as one of the possible future benefits of the project.

In Amsterdam, the municipality, along with the local energy provider and over eighty other public organizations, businesses and research institutes, is promoting the Amsterdam Smart City initiative. The scheme tests pilot projects linking ICT technologies to several urban dynamics, such as mobility, public-space use, business development and energy use. The overall aim is to reduce the city's carbon emissions, and the project places a strong emphasis on behavioural change. In its energy component, the scheme is testing innovative, small-scale projects on electric mobility, smart meters, smart appliances, in-home feedback displays and renewable-electricity generation. Those projects that show greater potential will be replicated at a larger scale.

For more information see: www.amsterdamsmartcity.nl/ and http://smartgridcity.xcelenergy.com/.

Andrés Luque, Department of Geography,
Durham University, UK

The drivers and challenges of mitigation

In their efforts to mitigate climate change, municipalities have employed a range of modes of governing, from a focus on self-governing actions such as reducing fossil-fuel use in their own vehicle fleets to regulating the use of private vehicles, and from the provision of low-carbon energy services to initiatives that seek to enable individuals to reduce their energy and water consumption. The extent to which municipalities have been able to develop and implement climate-mitigation policy and develop initiatives has been shaped by a range of drivers and challenges. In Chapter 4, three sets of factors – institutional, political and sociotechnical – were identified as providing both drivers and challenges for urban climate-change responses. The ways in which these factors shape climate-change mitigation vary across different urban contexts, but it is possible to identify key drivers and challenges in the urban-development, built-environment and urban-infrastructure sectors (Table 5.1).

As discussed in Chapter 4, *institutional* factors are those that shape the capacity of institutions to address climate-change mitigation. These include those that affect the co-ordination and conflicts between different levels of government and the extent to which partnerships with external organizations have been developed, as well as access to knowledge and financial resources. In all three sectors, the extent to which municipalities are able to intervene to promote climate-change mitigation depends upon whether they either own or have a stake in particular development projects, parts of the urban built environment or infrastructure network, and on their formal regulatory and planning powers or competency. In general, municipalities have limited authority in broad areas such as energy and transport policy, the development of urban infrastructures and services, the use of taxes or charges or in setting building standards, and more capacity in areas such as land-use planning and the development of voluntary programmes and initiatives (Bai 2007; Bulkeley and Betsill 2003; Collier 1997; Sugiyama and Takeuchi 2008; Schreurs 2008). In some cities, including in northern Europe, China, Latin America and Asia, municipalities can have more direct powers, either because these are mandated by central governments or because they have a stake in municipal energy, transport or waste companies, and in these cases their ability to intervene to address climate change is significantly higher (Bai 2007; Bulkeley and Kern 2006). However, even where these powers exist, significant challenges exist in terms of implementation and enforcement, ranging from explicit forms of corruption to the persistence of existing practices, even in the face of new regulation (Akinbami and Lawal 2009: 12; Bulkeley *et al.* 2009).

Table 5.1 Drivers and barriers for climate-change mitigation

	Urban development	Built environment	Urban infrastructure
Institutional drivers	Ownership of land/stake in development project. Supportive national and regional planning frameworks. Partnerships with proactive private-sector urban-development organizations. Access to sufficient capital to address additional upfront costs of eco-development projects	Own/operate housing and commercial stock. Supportive national and regional policy goals. Partnerships with private and civil-society organizations. Exchange of knowledge with other cities. Availability of external funding for initiatives. Flexible internal finance mechanisms to develop further projects	Own/operate infrastructure systems. Supportive national and regional policy goals. Partnerships with private and civil-society organizations. Exchange of knowledge with other cities. Access to capital for low-carbon infrastructure development. Flexible financing mechanisms to reinvest financial savings in further projects
Institutional challenges	Limited capacity and resources – knowledge, people, finance. Short-term payback periods. Determining and verifying the additional GHG-emissions reductions achieved through new projects. Absence of national or regional planning framework. Implementing and enforcing planning regulations. Lack of policy co-ordination and conflicting policy goals. Mismatch between urban jurisdiction and urban growth pressures	Limited capacity and resources – knowledge, people, finance. Short-term payback periods. Measuring and monitoring GHG-emissions reductions. Limited regulatory powers over existing building stock. Implementation/enforcement of regulation. Lack of policy co-ordination and conflicting policy goals	Limited capacity and resources – knowledge, people, finance. Short-term payback periods. Verifying impact of retrofit projects and additionality of new projects. Lack of municipal competencies in key infrastructure sectors. Lack of policy co-ordination and conflicting policy goals. Mismatch between urban jurisdiction and factors shaping development of urban infrastructure systems
Political drivers	Leadership – flagship projects, advancing city's international profile. Co-benefits for urban development agendas – e.g. economic opportunities, containing sprawl, meeting housing needs, climate	Leadership – opportunities to address concerns of key urban constituencies through e.g. improving housing, financial savings. Co-benefits – e.g. financial savings, addressing	Leadership – flagship projects that promote e.g. energy independence, modernization of transport systems. Co-benefits – e.g. financial savings, energy security, reducing air pollution, meeting basic needs, climate adaptation.

continued . . .

Table 5.1 ... continued

	Urban development	Built environment	Urban infrastructure
	adaptation. Windows of opportunity for large-scale redevelopment projects	poverty, enhancing building fabric, climate adaptation. Windows of opportunity for new buildings/interventions in existing buildings	Windows of opportunity for development of new infrastructure systems
Political challenges	Absence of leadership or political will. Mismatch between long-term process of urban development and short-term political cycles. Conflicts with agendas for urban development and growth	Absence of leadership or political will. Winning hearts and minds for behavioural-change schemes. Conflicts with other agendas for urban (re)development. Deliberate neglect of issues of energy affordability and access to decent living conditions	Absence of leadership or political will. Mismatch between long-term process of developing new infrastructure projects and short-term political cycles. Winning hearts and minds for behavioural-change schemes. Conflicts with other agendas to provide services and secure resources. Deliberate neglect of the basic needs of the marginal and vulnerable
Socio-technical drivers	Existing urban morphology that is conducive to compact city development. Emergence of niches/experiments for alternative technologies and social organization. New sociocultural expectations and practices for urban living that favour sustainability	Built environments that are conducive to forms of intervention to improve energy/water conservation. Emergence of niches/experiments for alternative technologies and social organization. New sociocultural expectations and practices for buildings that favour sustainability	Well-functioning and well-maintained infrastructure networks. Emergence of niches/experiments for alternative technologies and social organization. New sociocultural expectations and practices for energy, water and waste services that favour sustainability
Socio-technical challenges	Urban morphology based on historical legacy of sprawl or urban overcrowding, where compact development is impossible or inappropriate. Cultures of urban development based on continued availability of fossil fuels	Historical legacies that make retrofitting and redevelopment of buildings costly or challenging to existing urban aesthetics. Cultures of production and consumption based on continued availability of fossil fuels. Continuation of the 'predict-and-provide' mentality for service provision that excludes consideration of demand management	Infrastructure deficit and failure to meet basic needs. Rigid infrastructure networks and institutional cultures that favour incumbent technologies and prevent change. Continuation of the 'predict-and-provide' mentality for service provision that excludes consideration of demand management

Where municipalities lack direct powers to intervene in this way, the extent to which both vertical and horizontal forms of multilevel governance are in place that create enabling conditions for local action becomes even more critical (Chapter 4). An enabling, vertical multilevel-governance context, where municipal authorities have been supported in taking action by regional and national levels of government, can be a key factor in driving local responses. In countries as diverse as the UK, Sweden, Japan and China, national governments have provided planning frameworks, financial incentives and policy targets for municipalities for climate-change mitigation (Bulkeley 2009; Granberg and Elander 2007; Qi *et al.* 2008; Sugiyama and Takeuchi 2008). In Australia and the US, however, municipal responses to climate-change mitigation were developed in the late 1990s and early 2000s at precisely the time that national governments entrenched their opposition to taking action on the issue. In each case, municipal action was supported by the work of transnational, municipal climate-governance networks, sought inspiration from the wider international negotiation process, drew upon funding made available by federal governments and often took place in the context of supportive regional governments, illustrating that it is the multilevel-governance framework, rather than purely national-government support, that is important in fostering municipal responses. In particular, transnational, national and local networks can provide important opportunities for building urban climate-change-governance capacity. However, the impact of these initiatives has been uneven, with evidence suggesting that they have served to develop the capacity of those municipalities that are already leading responses to climate change, and that, although they are important vehicles for fostering political support and access to additional finance, 'in the absence of the financial and technological resources to execute programmes, the power of knowledge can be limited' (Gore *et al.* 2009: 22). In effect, networks and partnerships appear to be most important for those with a degree of existing capacity to act, leading to a virtuous circle where additional resources and support can be accessed (Kern and Bulkeley 2009).

These challenges of municipal powers and multilevel governance are exacerbated by the fragmented nature of urban governance. Particularly in large metropolitan areas, governance at the city scale may be shared by neighbouring municipalities or by overlapping levels of government. In London, for example, the Greater London Authority (GLA) and individual London borough councils all have a role in development planning. In Mexico City, research finds that:

> the administrative structure of city's governance differs from its boundaries and carbon-relevant socioeconomic and ecological functioning.

> Administratively, the city is managed by diverse federal, state and local tiers of government. Yet, the city functions as a complex system; its core area and localities, activities and households are interlinked by economic interchanges and transportation activities, by fluxes of materials and energy.
>
> (Romero-Lankao 2007: 529)

As this analysis makes clear, the drivers of GHG emissions, such as urban development, increased levels of mobility and continued use of fossil-fuel energy, are shaped by a complex set of public institutions, private companies and individual decision-makers, each of which is governed by a range of formal and informal institutions.

Access to additional forms of finance, through the European Union, national governments or other donors, can be one means through which such challenges are overcome, by providing the means to draw diverse actors together and to overcome existing budgetary constraints. Transnational municipal networks, such as the CCP programme and the C40, have been critical in leveraging funding for municipalities. Furthermore, the ability to establish novel financial mechanisms, whether internal 'revolving' funds, whereby the financial savings from, for example, energy-efficiency projects are reinvested in other mitigation activities, or forms of energy-performance contract, where third-party organizations take on the risks and responsibilities for achieving financial and GHG-emissions savings, has also proven to be a significant driver for urban mitigation action. Equally, a lack of access to finance, particularly in terms of the investment required in projects with longer payback periods than those that are regarded as standard, has proven to be a barrier to action. In cities where financial resources are so scarce that the basic needs of urban communities are not met, mitigating climate change is unlikely to attract, or indeed warrant, attention.

The *political* factors that shape urban responses to climate-change mitigation can be broadly considered in terms of issues of leadership, opportunity and conflict. Mitigating climate change has created significant potential for charismatic urban politicians to demonstrate their leadership on the issue of climate change, particularly in the face of seemingly endless international negotiations and national prevarication. The C40 network, for example, focuses on the ways in which 'cities act' and term themselves a 'climate leadership group'. Individual mayors also seek to promote the green credentials of their cities within such networks. Individual policy entrepreneurs within municipal authorities can also be important in terms of driving forward climate-change agendas. However, research suggests that such individuals may be a necessary, but not sufficient, factor in developing sustained climate-change-mitigation action, owing in part to the nature of the short-term political cycle of elections and because of the wider institutional and political barriers they encounter. In Durban, Mexico City and São Paulo, the effectiveness of political champions and policy entrepreneurs was found to be constrained by the wider contexts within which they operate (Aylett 2010; Romero-Lankao 2007; Setzer 2009). Windows of opportunity – specific events or moments in a particular urban context – can function as a means by which such barriers can be overcome, such as the opportunities afforded by the hosting of international sporting events for reconfiguring urban transport networks or undertaking specific urban-redevelopment projects. However, such effects can be short lived and may only serve as a means of intervening in specific parts of the city.

Most fundamentally, the political challenges of addressing climate change in the city stem from the ways in which the issue is regarded with respect to other key urban agendas. In some cases, municipal actors have been able to localize or reframe climate change, tying it to particular urban issues that are of significance – such as air pollution, congestion, the regeneration of urban areas – and have advanced the cause of mitigating climate change as a means of solving these more locally relevant issues. In other cases, the idea of addressing climate change, and in particular when it comes to reducing demand for those parts of urban life that many of us take for granted – water and energy use, transportation and the consumption of goods and services – has come into direct conflict with other urban agendas that seek to promote economic growth. Such challenges mean that it is particularly important to consider the urban contexts within which mitigation is being pursued. In developing economies, where resources are often more limited and other concerns are more pressing, 'subnational governments may be overloaded with other local demands, and climate policy may be down on the list of priorities' (Puppim de Oliveira 2009: 25). Whether or not climate-change mitigation should be pursued under such urban conditions, particularly if both total and per capita GHG emissions are low, is open for debate. In other, more affluent, urban contexts, the conflicts between addressing climate change and dominant urban political economies, often sustained by the very systems and practices that produce GHG emissions, may be more entrenched and serve to frame climate-change responses in a specific and narrow way. For example, in Portland, climate actions have been confined to:

> elements of energy consumption that could be influenced in an acceptable way by the municipal government. Energy used in flights to and from Portland International Airport, for instance, was excluded. Also excluded were the significant amounts of energy used in importing and exporting commodities, and the energy actually embodied in commodities.
> (Rutland and Aylett 2008: 636)

In part, these political challenges reflect the *sociotechnical* factors that also shape the possibilities and limitations for mitigating climate change in the city. The infrastructure systems, buildings and morphology of urban areas have been historically co-produced through networks of material infrastructures, institutions, interests and everyday practices that become self-sustaining (Figure 5.3). These sociotechnical networks tend towards obduracy – that is, to stability and the dominance of incumbent technologies and social interests (Hommels 2005). Where such networks are effective in providing services, such as shelter, energy provision and sanitation, incremental changes may be possible through developing new logics of efficiency and conservation that, in turn, serve to mitigate climate change. Where such networks are fragmented, in disrepair or simply inadequate, their continued presence in the urban landscape serves to create an infrastructure deficit that sustains inequalities and actively excludes poorer and more marginalized communities. In both cases, more radical change requires that existing networks are challenged and disrupted, in order that new forms of, in this case, low-carbon network can be developed. As discussed above, in some cases, climate-mitigation agendas are giving rise to the emergence of alternative forms of service provision that take place beside, or in the place of, dominant forms of energy or water provision. For the most part, these have taken the form of niche or experimental networks and practices (Chapter 7). Although they may offer alternatives to existing forms of provision, the extent to which they can be developed at sufficient scale may be limited.

Conclusions

Mitigation has been the issue on which urban responses to climate change have focused. Many municipalities have developed elaborate policy approaches, based on measuring and monitoring their emissions, setting targets and developing action plans. Although the extent to which such policy approaches have been realized in practice is debatable, the willingness and enthusiasm with which municipalities have embraced such approaches are testament to the way in which climate change has now entered urban policy agendas. Furthermore, the lack of a coherent policy framework does not mean that actions to address climate-change mitigation have not been forthcoming. Across the urban-development, built-environment and urban-infrastructure sectors, a range of actions and activities can be pinpointed that have been undertaken in response to climate change. Municipalities have engaged in a variety of self-governing and enabling actions, and in some cases have also developed forms of provision and regulation in order to mitigate climate change. However, there is limited evidence about the effect of individual

Figure 5.3 Existing air-conditioning systems and norms about indoor thermal comfort in Hong Kong may limit the potential for energy efficiency
Source: Harriet Bulkeley

initiatives, and, as a result, it is difficult to determine what this all adds up to and the extent to which it is making a difference, either in terms of absolute or relative GHG-emissions reductions or in terms of reconfiguring urban development along low-carbon pathways.

The extent and nature of climate-change mitigation in particular urban contexts have been shaped by the highly differentiated levels of institutional capacity that exist between cities, as well as by the political and sociotechnical challenges that have been encountered. Where action has been forthcoming, critical drivers have included the extent of municipal competencies, the multilevel-governance context, the availability of finance, as well as the presence of political will and opportunities to engage with, and transform, existing sociotechnical networks. Where gaps between rhetorical commitments to climate change and action on the ground persist, the lack of an enabling governance context, limited resources, political conflicts and obdurate sociotechnical infrastructures have been the most significant factors limiting action. In this context, it is important to recognize that, although efforts to enhance institutional capacity – through, for example, developing more knowledge about the potential to reduce GHG emissions locally, creating additional sources of finance, improving policy co-ordination and developing new forms of information sharing and exchange – are important, they will not be sufficient to address more fundamental political and sociotechnical challenges facing urban climate-change mitigation. Addressing these challenges requires the forging of new forms of political alliance, between municipalities and other key urban actors who critically shape urban political economies, as well as the development of interventions and alternatives within existing sociotechnical networks (Chapter 7).

The resulting landscape of urban responses to climate-change mitigation can best be summarized as a patchwork, where many thousands of cities globally have declared their intentions to mitigate climate change, but the resulting actions and emissions reductions are rather more difficult to discern. Although it is, therefore, clear that cities are now firmly part of the international effort to address climate change, evaluating the specific contribution that cities are making remains challenging at both the level of individual initiatives and as a whole (Chapter 8).

Discussion points

- Why have municipal responses to climate change historically focused on mitigation rather than adaptation? What are the implications of this policy focus for the future of urban climate-change policy?

- For one or more case study cities, compare and contrast the strategies and measures that are being undertaken in the three main policy sectors – urban development, built environment and urban infrastructure – discussed in this chapter. In which sectors has policy been most fully developed and implemented? What factors may explain the differences that you find?
- How far can we undertake a comparative analysis of the drivers and challenges of mitigating climate change in different urban contexts? Are there some factors that 'travel' between urban contexts, or are the histories and geographies of individual cities such that such comparisons are neither useful nor necessary?

Further reading and resources

Given that the focus of urban responses to climate change has been on mitigation, most of the research in this field has focused upon this issue. All of the core books listed in Chapter 1 contain detailed discussion and examples of municipal climate-change-mitigation policy. In addition, the following articles are just some examples of the useful case studies that are now available:

Gustavsson, E., Elander, I. and Lundmark, M. (2009) Multilevel governance, networking cities and the geography of climate-change mitigation: two Swedish examples, *Environment and Planning C: Government and Policy*, 27: 59–74.

Puppim de Oliveira, . (2009) The implementation of climate change related policies at the subnational level: an analysis of three countries, *Habitat International*, 33(3): 253–9.

Romero-Lankao, P. (2007) How do local governments in Mexico City manage global warming? *Local Environment*, 12(5): 519–35.

Rutland, T. and Aylett, A. (2008) The work of policy: actor networks, governmentality, and local action on climate change in Portland, Oregon, *Environment and Planning D: Society and Space*, 26(4): 627–46.

Evidence of the sorts of policy and strategy that are being developed at a local level can be found at the following websites from municipal networks, as well as by searching online information about individual cities:

- Climate Alliance: www.klimabuendnis.org/member_activities0.html
- Covenant of Mayors: www.eumayors.eu/index_en.html
- ICLEI CCP: www.iclei.org/index.php?id=800 (and links to regional campaigns).

In addition, the Global Carbon Project has a dedicated website that collects examples of urban and regional responses to climate-change mitigation and hosts a range of different materials:

- www.gcp-urcm.org/Resources/HomePage.

6 Urban adaptation – towards climate-resilient cities?

> The message is clear. We need to put a system in place to have tangible projects, working with local government, community-based organizations, and non-government organizations. We need to avoid seminars and go directly to doing things, and we must have a bottom-up approach. And these activities must take place yesterday, because tomorrow is too late.
> (Mayor Adam Kimbisa, Dar es Salaam, 18 June 2009, quoted in Dodman *et al.* 2011: 13)

Introduction

As we saw in the last chapter, mitigation, or reducing GHG emissions, is a critical response to climate change. There is, however, growing evidence that some form of climate change will be the inevitable result of the legacy of the GHG emissions that have been produced over the past century, and, in response, there has been a growing call for societies to be prepared to adapt to the consequent changing climatic conditions. As Mark Pelling (2011a: 6) argues, adapting to changing environmental conditions is 'nothing new. Individuals and socio-ecological systems have always responded to external pressures'. What makes adapting to climate change particularly challenging is uncertainty about the sorts of risk that might be experienced, the diffuse ways in which changing climates might impact on particular places at particular times, and the distant nature of how the issue is experienced on a daily basis, particularly in comparison with more urgent forms of risk and disaster (Pelling 2011a). As a result, 'a focus on adaptation now may be perceived as an acknowledgement of allocating scarce public resources to a threat that is not (yet) perceived as imminent' (Laukkonen *et al.* 2009: 289).

It is partly because of these facets of uncertainty, diffuse impacts and the seemingly distant nature of the problem that adaptation has, to date, received little attention from policy-making communities at the international, national and local levels, compared with mitigation. At the same time, the political necessity of demonstrating that the international community was prepared to take action to reduce future climate change also served to reduce the emphasis placed on adaptation. Adaptation, some suggest, is also a more challenging political and policy problem. Whereas 'mitigation is a bounded problem', where progress against targets can be assessed by measuring levels of GHG emissions and atmospheric concentrations, adaptation 'is messier, concerned with adjustments in human systems at different scales (local to global) and by different actors (e.g. government, individuals, households, etc.) and which may only be partially developed in response' to climate change itself and which is very difficult to measure (Berrang-Ford *et al.* 2011: 26).

The combined challenges of adaptation as an issue and the dominant concern of international and national policy-making communities with mitigating climate change over the past two decades have meant that efforts to develop and implement adaptation strategies at the urban level are only just beginning. Despite the considerable vulnerability of cities to the effects of climate change, and the long history of climate-related disasters that have affected urban areas, this has been exacerbated by the fact that 'cities . . . have received much less attention than rural areas in adaptation literature and practice' (Commission on Climate Change and Development 2009: 25). As the mayor of Mombassa so clearly puts it in the quote that opens this chapter, the continued neglect of urban adaptation is a significant concern because of the urgent need for action to address current and emerging vulnerabilities in the context of climate change.

Despite this unpromising context, it is possible to identify the initial stages of what could be termed urban climate-change adaptation. The rest of this chapter is divided into four sections. The first section explores just what adaptation is, examining some of the key terms and ideas that are used to describe adaptation. The second section considers how adaptation policy is emerging in different urban contexts and the differences emerging between government-led and community-based adaptation approaches. The third section turns to assessing the ways in which adaptation is emerging in practice in three sectors – urban development, built environment and urban infrastructure. The drivers and barriers that are shaping the emergence of adaptation in urban contexts are then examined in the fourth section. The conclusions summarize the main findings of the chapter and provide some discussion questions and further reading and resources to investigate.

What is adaptation?

Adaptation is a 'deceptively simple concept' (Pelling 2011a: 20), with a common-sense meaning – a process of response and change – that has acquired specific definitions within the lexicon of terms used to describe and analyse the climate-change phenomenon. The IPCC, the leading scientific authority on climate change, describes adaptation as 'adjustment in natural or human systems in response to actual or expected climatic stimuli or their effects, which moderates harm or exploits beneficial opportunities' (IPCC 2007e; see Box 6.1). Behind this straightforward approach are, however, central questions about the change that is being responded to, who is doing the adapting and why, how this is achieved, and where the limits to adaptation lie (Pelling 2011a: 13). As Chapter 2 sets out, much of the focus on vulnerability at the urban scale has focused, on the one hand, on the gradual and long-term issue of sea-level rise and, on the other hand, on current climatically induced disasters, such as storms and droughts. Whether adaptation can simultaneously address long-term incremental change, current climatic events and potential future abrupt climate change is a matter of significant debate. Furthermore, vulnerability is also underpinned by other social and economic conditions in the city (Chapter 2). Questions, therefore, emerge as to whether climate-change adaptation involves only responses to climatic stimuli or must also entail responding to wider processes that shape vulnerability. This, in turn, relates to who undertakes adaptation, and whether this is done in a planned, purposeful manner through processes of policy-making or the organization of private and community interests, or is conducted autonomously by individuals within a society (Box 6.1).

Once the issues and actors with which adaptation should be engaged have been identified, there are multiple means by which it can be undertaken, each of which encounters specific challenges and limitations. One important distinction in the literature is between adaptation and coping. Work in the adaptation field has partly been derived from a longer tradition of research about how households and communities respond to disasters and how disaster risks can be reduced. Studies have shown how risk is reduced through different coping strategies. In the climate-change-adaptation field, coping is used to refer to:

> existing strategies that are used by urban residents to respond to climate variability and other threats. These are often short-term efforts to prevent injury or damage to property; usually require little financial expenditure; and are not part of a larger or more structured plan. In addition, they tend

to address immediate symptoms rather than the root causes of vulnerability; and rarely result in longer-term resilience or greater levels of sustainability.

(Dodman *et al.* 2011: 6)

In effect, coping can involve multiple different means by which households and communities seek to draw on their existing social and economic resources to survive in the face of immediate risk, and by which they seek to repair and rebuild in the aftermath of such events. Although the term implies a benign process, the reality is more stark, for 'often acts labelled as coping require the expenditure or conversion of valuable assets to achieve lower order outcomes, undermining current capacities and future development options' (Pelling 2011a: 34). In contrast to coping, adaptation can be regarded as a more complex and deliberate process, 'through which an actor is able to reflect upon and enact change in those practices and underlying institutions that

---

BOX 6.1

What is adaptation?

> Adaptation is . . . adjustment in natural or human systems in response to actual or expected climatic stimuli or their effects, which moderates harm or exploits beneficial opportunities. Various types of adaptation can be distinguished, including anticipatory, autonomous and planned adaptation:
>
> **Anticipatory adaptation** – Adaptation that takes place before impacts of climate change are observed. Also referred to as proactive adaptation.
>
> **Autonomous adaptation** – Adaptation that does not constitute a conscious response to climatic stimuli but is triggered by ecological changes in natural systems and by market or **welfare** changes in **human systems**. Also referred to as spontaneous adaptation.
>
> **Planned adaptation** – Adaptation that is the result of a deliberate policy decision, based on an awareness that conditions have changed or are about to change and that action is required to return to, maintain, or achieve a desired state.
>
> (IPCC 2007e)

generate root and proximate causes of risk, frame capacity to cope and further rounds of adaptation to climate change' (Pelling 2011a: 21). This definition suggests that, for responses to 'climatic stimuli' to count as adaptation, they must be of a particular kind: they must be considered and deliberate. Simply adjusting to the onset of climate-related disasters by, for example, moving to higher ground or selling family assets may demonstrate the capacity to cope but is not sufficient to be considered 'adaptation', which would require some form of premeditation of what needed to be done and by whom in order to change practices and institutions to reduce risk.

The ability to make such changes is often referred to as adaptive capacity. As we saw in Chapter 2, vulnerability to climate risks is regarded as a function of the exposure to risk, the susceptibility of particular systems and individuals to that risk, and their adaptive capacity. Through the deployment of different strategies and measures, adaptation can work to reduce risk, either by reducing exposure or by reducing the impacts of that exposure. Adaptive capacity is thought to vary significantly within and between households, communities and places, as well as over time, and to be shaped by both social factors (such as health, education, social capital) and physical attributes (including the material conditions in which people live, technology and material wealth) (Adger *et al.* 2005; Eakin and Lemos 2006; Hobson and Neimeyer 2011). Of critical importance for understanding the capacity to adapt in urban settings is what has been termed the 'adaptation deficit' – the *lack* of basic infrastructure provision. Infrastructure systems, which provide water, sanitation, shelter, mobility, energy and so on, are critical in shaping access to basic services, employment opportunities, health and other critical attributes that reduce vulnerability. In many urban contexts, it is this deficit of infrastructure and services that is the critical determinant of adaptation capacity, for 'you cannot climate-proof infrastructure that is not there', and investment in addressing this deficit is likely to be of limited utility if 'there is no local capacity to design, implement and maintain the necessary adaptation measures' (Satterthwaite *et al.* 2008b: 9). The significance of the adaptation deficit for many urban places points to the ways in which adaptive capacity is not only shaped through the attributes of individuals and households, but is also structurally related to historical and contemporary processes of urban development.

Adaptation, as a deliberate and considered process of responding to climatic conditions in the context of wider social, economic and political processes, is therefore shaped by the interaction between vulnerability, adaptive capacity and adaptation deficits. Further complexity arises when we consider how and to what ends adaptation might be undertaken – what is it that adaptation should be seeking to achieve? Within policy-making and academic communities, there is a growing interest in the notion of resilience as the basis for achieving adaptation, in part because it is seen as a term that can also be used to advance mitigation agendas (Leichenko 2011: 164; Box 6.2). The concept of resilience has long intellectual roots in the study of ecological systems, but, within the 'broad array of urban resilience literatures', it is 'typically understood as the ability of a system to withstand a major shock and maintain or quickly return to normal function' (Leichenko 2011: 164). Rather than seeking to prevent or avoid risk, the term resilience has been used as a means of signifying a form of adaptation that recognizes the need to live with risk and to foster social and organizational learning about how such responses can be undertaken (López-Marrero and Tschakert 2011: 229; Pelling 2011a).

However, despite these broadly common concerns, there is significant disagreement about the 'characteristics that define resilience and the appropriate analytical unit for the measurement of resilience' (Leichenko 2011: 164). In its most straightforward interpretation, resilience has been considered as the 'opposite of vulnerability. The more resilient, the less vulnerable', but, as Pelling (2011a: 42) goes on to argue, 'this belies the complexity of the conceptual relationship between these two terms.' Resilience can involve 'resistance and maintenance', where political authorities seek to resist the notion that fundamental or significant change is required and maintain existing systems; 'change at the margins', where risk is acknowledged, and some of the symptoms of risk are addressed, insofar as they do not threaten existing orders; or 'openness and adaptability', where the root causes of risk are addressed in a flexible manner that is able to address uncertainty (Pelling 2011a: 43–4, after Handmer and Dovers 1996). Despite the different ways in which the term can be used, within the climate change area it is the narrower

---

BOX 6.2

Defining resilience

> The ability of a social or ecological system to absorb disturbances while retaining the same basic structure and ways of functioning, the capacity for self-organization and the capacity to adapt to stress and change.
> (IPCC 2007f)

version of resilience that is most commonly used. The IPCC, for example, refers to resilience as a system that can 'absorb disturbance' and retain 'structure' while maintaining its ability to adapt to stress and change (Box 6.2). For Pelling (2011a: 55), this suggests that resilience can be considered as one form of adaptation that 'seeks to secure the continuation of a desired system's functions into the future in the face of changing context, through enabling alteration in institutions and organisational form'.

Resilience can therefore be considered as one possible form of adaptation response. It moves beyond 'coping' by involving some form of deliberate intervention or response, but essentially serves to sustain existing systems and practices (Table 6.1). Mark Pelling identifies two additional forms of adaptation that go beyond resilience to address the more structural causes of vulnerability. Transitional adaptation 'is targeted at reform in the application of governance' (2011a: 69), recognizing the fundamental role that forms of governance have in structuring the ways in which problems are framed, solutions are implemented, and rights and responsibilities are shared. For Pelling, 'opportunities for transition arise when adaptations, or efforts to build adaptive capacity, intervene in relationships between individual political actors and the institutional architecture that structures governance regimes' (2011a: 82). A third level of adaptation may emerge, where it operates as a 'mechanism for progressive and transformational change that shifts the balance of political or cultural power in a society' (Pelling 2011a: 84). Rather than being concerned with the 'proximate causes' of adaptation, such as infrastructures, livelihood planning and so on, transformation is 'concerned with the wider and less easily visible root causes of vulnerability' (Pelling 2011a: 86; Table 6.1). These levels of adaptation are regarded by Pelling as 'nested and compounding', so that changes at one level may create opportunities at another. Equally, he recognizes that, 'on the ground mosaics of adaptation are generated from the outcomes of overlapping efforts to build (and resist) resilience, transition and local transformative change and remaining unmet vulnerabilities' and shift over time in response to dynamics of risk and adaptive capacity (Pelling 2011a: 24).

Pelling's layered account of adaptation is useful because it helps us to think through the different ways in which adaptation might be pursued, and how this is orchestrated by different social, economic and political objectives. On the whole, adaptation has been interpreted by international organizations and national governments as a form of 'expert-based risk management', and typical responses have included 'increasing the robustness of infrastructures, enhancing the protective functions of ecosystems, incorporating climatic risks in development planning, market solutions, establishing emergency

Table 6.1 Levels of adaptation

	Resilience	Transition	Transformation
Concept	Adjustment to practices and institutions to improve performance without changing guiding assumptions, routines and norms	Incremental changes to practices and institutions in the light of reflection on existing framing of problems and goals	Irreversible change in the regime of practices and institutions in order to create new possibilities for action and socio-ecological change
Goal of adaptation under each approach	Functional persistence in a changing environment	Realize full potential through the exercise of rights within the established regime	Reconfigure the structures of development
Scope of adaptation in each approach	Change in technology, management practice and organization	Changes in the practices of governance to secure procedural justice (who can participate and what rights they have over decision-making), which can lead to incremental change in the governance system	Change overarching political–economic regime
Examples	Resilient building practices. Development of new seed varieties	Implementation of legal responsibilities by private- and public-sector actors and exercise of legal rights by citizens. Community-based schemes that recognize and address the needs of the marginalized	New political discourses that redefine the basis for distributing security and opportunity in society and socio-ecological relationships

Source: Adapted from Pelling 2011a: 23–4, 51

funds, improving societal awareness and preparedness, reducing institutional fragmentation, and creating policy frameworks for disaster management' (Manuel-Navarrete *et al.* 2011: 249). Pelling's threefold framework suggests that, although such approaches may be capable of enhancing resilience, they do little to address or challenge the underlying social, economic or political basis of vulnerability. Others, notably academics and NGOs, have sought to challenge this dominant model by advocating an approach that focuses on reducing the vulnerability of the poor and marginal in society, through changing governance structures to enhance participation and 'building capacities for self-protection and group action ... community risk assessments ... revaluing traditional coping practices ... and mobilizing social capital' (Manuel-Navarrete *et al.* 2011: 250). Although such approaches begin to consider how and why existing practices and institutions may need to change, they are usually focused on forms of *transitional* adaptation rather than radical transformation. In contrast, a 'critical adaptation agenda sees the experiencing of hazards as essentially political and tied to contingent development paths' (Manuel-Navarrete *et al.* 2011: 250). Understanding the limits and opportunities for adaptation, therefore, entails examining 'the conflicts arising between visions of development, and how these conflicts create possibilities for altering development paths, transforming governance structures, and generating coping strategies' (Manuel-Navarrete *et al.* 2011: 250).

Making adaptation policy

Given the multifaceted nature of adaptation and the ways in which vulnerability to climate change is deeply embedded within other forms of urban vulnerability, identifying just what counts as climate-change adaptation is challenging. At the same time, as the introduction to this chapter made clear, the need for cities to address climate-change adaptation has only recently been realized as an explicit policy goal. It is within this context, of an amorphous issue area that is at a nascent stage of development, that processes of making climate-adaptation policy are taking place in cities. Furthermore, there is enormous variability in the urban conditions, contexts and processes that shape the possibilities and limits for adaptation – the adaptation deficit and adaptive capacity of individuals, communities and urban systems vary significantly across and between urban places.

Of particular importance are the differences that can be found between cities in less economically developed countries and those in more economically developed countries. For many of the former, issues of adaptation deficit are central, as, for a large section of the urban population in developing countries, little can be expected of local and national governments as they currently lack the capacity or willingness to provide the basic infrastructure and services that are central to adaptation.

(UN-Habitat 2011: 131)

In such cities, climate-change risks will serve to exacerbate existing vulnerabilities. In contrast, although, in developed economies, practices and institutions designed to cope with environmental risk are almost universally established – including access to basic services, social capital, finance, insurance, state assistance and so on – 'this does not mean that adaptation is necessarily given the priority it deserves. There are many relatively wealthy cities that need major upgrades in their infrastructure that should take account of climate change impacts' (UN-Habitat 2011: 142). In this case, it is the additional risks that climate change poses that are often overlooked and for which existing practices and institutions are seen as insufficient.

Demonstrating just how ingrained such differences may be, a recent review of the literature in the adaptation domain finds 'distinct profiles' of how adaptation is taking place in different national settings. In low-income countries, adaptation is generally characterized as reactive or responsive, based on individuals, with 'weak involvement of government stakeholders', most likely in the natural resource sector, and 'adaptation mechanisms are more likely to include community-level mobilization rather than institutional, governmental or policy tools' (Berrang-Ford *et al.* 2011: 31). In high-income countries, adaptation, although still low on the policy agenda, is more 'proactive or anticipatory', 'more likely to include governmental participation', to focus on non-resource sectors, and to involve longer-term planning activities, 'such as preparation for projected impacts, monitoring, increasing awareness, building partnerships, and enhanced learning or research' (Berrang-Ford *et al.* 2011: 31). Although such broad distinctions, between north and south, high and low income, inevitably leave many of the differences between urban places unexplored, such findings are useful in pointing to the very different urban contexts within which adaptation policy is emerging.

It is in these contexts that questions of who is adapting to climate change, why and how, become important. In the remainder of this section, the emergence and development of municipal adaptation policies are discussed, before we turn to consider alternative approaches that have placed communities at the heart of adaptation responses. Although some see municipal governments as having an essential role to play in providing the regulatory context and co-ordinating action needed to adapt at the urban level, for others, such actors can lack the capacity, political will or interest in adaptation, which instead is regarded as a matter for communities to address directly. Given the significant challenges of adaptation, and the need for households and communities to be engaged in these processes in order to realize the benefits of such strategies and measures, some level of community involvement is important in adaptation. However, community-based responses can rarely provide a substitute for the longer-term, more investment-intensive and systemic changes required to, for example, develop new infrastructure systems, and they are often disempowered from being able to make fundamental changes to existing political and economic systems. In this context, it is important to ask why and how communities are being engaged with the adaptation agenda, to consider whose interests are best served through such mechanisms, in order to evaluate their utility and value.

Adaptation and municipal planning: towards integration?

Municipal responses to climate-change-adaptation planning have tended to follow a similar set of steps to those involved in mitigation of assessment, goal setting and implementation (Chapter 5). Initial work has focused on defining and assessing the potential climate-change risks, before developing strategies and then undertaking implementation and evaluation. In the absence of specific protocols or standards for developing adaptation responses at the urban level, research suggests that, 'a number of cities have thus mimicked the sequential and inventory-based approach taken to mitigation by initiating their adaptation processes with risk and vulnerability assessments' (Anguelovski and Carmin 2011: 170). Examining each of these stages in turn can provide insights into how, and with what effect, adaptation responses are being developed at the urban scale.

As UN-Habitat notes in its 2011 report on cities and climate change, 'generally, the first evidence of an interest by city or municipal governments in climate change is an interest in assessing the scale and nature of likely risks' (UN-Habitat 2011: 138). As discussed in Chapter 2, assessing climate risks and vulnerability at the urban scale is beset by challenges of downscaling existing models of predicted climate change to meaningful scales for urban decision-makers. The challenges of assessing the level of risk and vulnerability to climate impacts go beyond the complex issue of creating relevant climate science at the urban level. As is also the case with mitigation, there are also multiple problems, particularly for cities in developing countries, regarding the extent to which urban authorities have access to sufficient data, for example about critical urban characteristics (such as existing infrastructure networks, population, health and medical needs etc.) or the nature and extent of existing urban settlements upon which to base their predictions of the impacts of climate change. As a result, 'climate impact forecasts regarding extreme events and respective adaptation strategies hardly exist on the local scale' (Birkmann and von Teichman 2010: 175). In a few cases, municipal authorities have made specific efforts to overcome this lack of data and to consider current and future climate risks in the city, examples of which include Cape Town, Ho Chi Minh City, London, Halifax and Boston (Birkman *et al.* 2011). Such efforts have been particularly extensive in London, where:

> The Greater London Authority have analysed how London is vulnerable to weather related risks today (and so established a baseline to assess how these risks change) and then used climate projections from climate models to identify how climate change accentuates existing risks and creates new risks, or opportunities in the future. This enables the GLA to assess and prioritise the key climate risks to London.
>
> (Nickson 2011: 4)

In seeking to identify risks and to build resilience to particular climate-adaptation challenges, several cities have developed adaptation strategies that focus on specific sectors (e.g. coastal management, health/heat plans) (Anguelovski and Carmin 2011: 170). In Europe, evidence suggests that specific plans for adaptation are often part of wider climate-change and sustainability strategies, where concerns for mitigation remain dominant, as is the case, for example, in Manchester and Madrid (Carter 2011: 195). Other cities, including London, Copenhagen and Rotterdam, have preferred to develop 'stand-alone' adaptation strategies (Carter 2011: 195). On the one hand, such strategies can raise the profile of adaptation as an issue in its own right, but, on the other, they can serve to leave adaptation concerns on the sidelines when it comes to mainstream urban policy-making. Recognizing this challenge, some municipal authorities have sought to establish processes and structures through which adaptation concerns are integrated into decision-making frameworks across different sectors. In the north-west United States, King County, the area within which Seattle is located, was an early pioneer in establishing an interdepartmental climate-change team in 2006, in order to build 'scientific expertise within their county departments to ensure climate change was considered in future policy, planning, and capital investment decisions' (Pew Centre 2011: 17). In Seattle itself, the 2006 Climate Action Plan called for the formation of a similar team to address climate adaptation in several areas, including sea-level rise, storm-water management, urban forestry and heatwaves, and, in 2008, eighteen government departments were tasked with analysing future vulnerabilities and the scope for strategies to address climate impacts in critical areas of service delivery and in order to protect public assets (Pew Centre 2011: 19). Nonetheless, across North America, such strategies remain few and far between, and adaptation policy remains at an early stage of development (Zimmerman and Faris 2011; see the city case study on Philadelphia).

CITY CASE STUDY

Adapting to climate change in Philadelphia

Philadelphia is the urban centre for a greater metropolitan region of approximately 6.3 million people, living within the Greater Philadelphia region comprising twelve counties in four states on the eastern seaboard of the USA. It was founded in the late seventeenth century and, in the early twentieth century, became a powerhouse of America's industrial growth. Over the second half of the twentieth century, like many other cities in developed economies, Philadelphia entered a period in which de-industrialization and decentralization drove processes of urban decay as factories closed, population declined and land lay vacant. Like many American cities, problems of urban degeneration and poverty are most acute within the predominantly African American neighbourhoods that lie within the city limits. Nonetheless, the urban population has recently started to increase through a process of regeneration.

It is in this context of the uneven impacts of economic decline and regeneration that the impacts of climate change are taking place. A key risk of climate change in Philadelphia is increased hot weather in summer and more frequent and prolonged heatwaves. Climate change has resulted in the average annual temperature in the north-east USA increasing by 2°F since 1970. This has resulted in temperatures that more frequently reach 90°F (32.2°C), and, over the next several decades, temperatures are predicted to rise an additional 2.5–4°F in winter and 1.5–3.5°F in summer (Karl et al. 2009). Under the higher-emissions scenarios of the IPCC, Philadelphia would experience in excess of 80 days per year over 90°F (32.2°C) and close to 25 days over 100°F (37.8°C) by late this century. Summer conditions would be extended by 6 weeks, and heatwaves would become more common (Karl et al. 2009).

The impact of these changes is exacerbated in Philadelphia because of the phenomenon known as the urban heat island effect, in which the high proportion of impervious and heat-absorbing materials in the city (particularly concrete, asphalt, and bricks in buildings and roads) results in air temperatures significantly higher than in the surrounding countryside. This combines with high concentrations of vulnerable people, such as the elderly and young children, the unwell and the poor, who live in cities to make heat the most significant cause of weather-related morbidity in the USA. In fact, Philadelphia has been labelled the 'Heat-Death Capital of the World' (EPA 2006, quoted in Union of Concerned Scientists 2008: 13). For instance, a four-day heatwave in July 2008 was responsible for eight deaths. Its particular vulnerability is produced through the combination of ageing housing stock and acute poverty. Over 90 per cent of residential housing in Philadelphia is terraces – referred to as 'rowhomes'. These homes were rapidly constructed to meet the city's early period of growth and were typically built with minimal or no insulation, and their almost-flat roofs were waterproofed with black asphalt, which absorbs solar energy, transmitting it as heat into the rooms below through uninsulated attic/loft spaces. Many of them are now over 100 years old, and one in five residents lives below the poverty line. At the same time, these dwellings are often poorly sealed against draughts, meaning that they can be very cold in winter.

A second climate change-related risk in Philadelphia is the provision of fresh water for both drinking and industrial use. Philadelphia draws most of its water from the Delaware River at a point just upstream from where the fresh water flowing down the Delaware River mixes with the salty water from Delaware Bay. Sea-level rise might have a similar impact to droughts, in which saline water from the Delaware Bay extends further north in the Delaware estuary and river system, jeopardizing the quality of the city's water supply (Union of Concerned Scientists 2008: 16). This could be compounded by reduced flows of fresh water down the Delaware River as a result of reduced winter snowfall further upstream. At present, average chloride concentration at the Philadelphia Water Department's intake is around 21 mg/l. Health impacts among vulnerable people start to manifest themselves when concentrations exceed about 50mg/l (Union of Concerned Scientists 2008: 16). Climate change could also increase the incidence of heavy rainfall events, as well as causing more winter precipitation to fall as rain rather than snow (Karl *et al.* 2009). Such intense rainfall events are particularly problematic in Philadelphia because it has a combined sewer system that carries both storm-water and sewage to treatment plants through common pipes. This means that, if the capacity of the treatment plants is exceeded, the excess water and sewage flows into waterways untreated, compromising water quality and potentially contaminating drinking-water supplies (Union of Concerned Scientists 2008: 16).

It has been in the context of these combined climate-change risks and the leadership of local politicians and officials that the municipality has started to respond to both mitigation and adaptation. Released in 2007, the *Local Action Plan for Climate Change* was Philadelphia's first climate-change plan (City of Philadelphia 2007). It built on the city's commitment to three GHG-reduction initiatives: ICLEI's CCP Initiative, the US Mayors' Climate Protection Agreement and the Large Cities Climate Leadership Group and Clinton Climate Initiative. In 2009, action on climate change and urban redevelopment was accelerated with the City of Philadelphia's *Greenworks Philadelphia* plan (City of Philadelphia 2009), which converted many of the aspirational goals of the earlier *Local Action Plan* into more concrete targets, with a delivery date of 2015. *Greenworks* was organized into five themes – energy, environment, equity, economy and engagement – and fifteen targets. It aimed to address the key climate-change risks facing the city through a commitment to urban sustainability:

> Philadelphians understand why this work is important. They know that the Mayor's call for Philadelphia to become the 'greenest city in America' is not just about preventing ice caps from melting or crops from drying up thousands of miles away, but also about decreasing the cost of cooling a Southwark house in the summer or heating it in the winter; reducing the number of trips a mother in Oak Lane takes to the hospital with her asthmatic son; preventing sewage from backing up into a basement in Northern Liberties; and giving every child in every neighborhood a safe, clean, healthy place to play.
> (City of Philadelphia 2009: 3)

The most notable climate-change adaptation, which pre-dates *Greenworks*, is Philadelphia's Heat Health Watch Warning System, which was launched in July 1993 after a heatwave that killed over 100 people. The first such system to be implemented in the USA, Philadelphia's Heat Health Watch Warning System issues a heat alert and follows this up with targeted visits by the health department and block captains to vulnerable people, such as the elderly and homeless. At the same time, electric utilities refrain from shutting off services for nonpayment, and public cooling places extend their hours. These measures are complemented by a 'heatline' that people can call if they are experiencing health problems. Nurses staff the line and can dispatch mobile units to the residence if necessary (Union of Concerned Scientists 2008; Karl *et al.* 2009). The system was estimated to have saved 117 lives in its first 3 years (Karl *et al.* 2009, p. 91), along with a 'Cool Homes Program', which provides measures such as insulation and roof coatings to elderly, low-income residents, both to save energy and to increase their comfort on hot days. One weakness of the programme in the context of climate change is its dependence on air-conditioning use as a means of coping with heatwaves, as this is very energy-intensive.

This weakness is, to some extent, being addressed through a second adaptive mechanism – Philadelphia's *EnergyWorks* programme, a low-interest revolving-loan programme that provides finance for energy-efficiency works. It was designed to achieve the second goal of *Greenworks*: namely, to reduce citywide energy consumption by 10 per cent. A major challenge for Philadelphia in adapting to climate change is improving and upgrading ageing residential buildings, so that homes are more energy efficient and more comfortable for those living in them. *EnergyWorks* provides 0.99 per cent loans of up to US$15,000 for energy-efficiency improvements or 'retrofits', such as installing windows and doors, upgrading heating and cooling systems and installing insulation. It is designed to help residents cope with hotter summers without increasing their energy consumption, and also to help reduce the amount of energy used for air conditioning, most of which still comes from GHG-emitting fossil fuels. However, the fact that it is a loan programme means that many potential beneficiaries are unable to take advantage of it because of poor or bad credit histories. A separate programme is in place that provides similar upgrades to low-income residents, but it is characterized by long waiting lists.

The EPA has had a policy mandating that cities eliminate or substantially reduce combined sewer overflow since 1994, but compliance has been a problem

Figure 6.1 Residents of rowhomes in Philadelphia may be particularly vulnerable to increased incidents of heatwaves
Source: Gareth Edwards

(Karl et al. 2009: 94–5). As part of Philadelphia's *Greenworks* plan, and in recognition that the problem will only get worse with climate change, in 2009 the Philadelphia Water Department launched a plan called *Green City, Clean Waters*, which aims, in the next 25 years, to replace a third of the city's impervious surfaces with green infrastructure that can intercept storm-water and allow it to infiltrate into the ground, reducing both flooding and the health effects of sewer overflows.

In Philadelphia, the challenge of responding to climate change is seen both as one of mitigation and as one of adaptation, and the case study illustrates how these issues are often linked together – higher temperatures leading to the increased use of energy in efforts to cool indoor space. Reducing energy use, while also adapting to climate change, requires the sorts of integrated approach that are now beginning to be implemented in the city.

Gareth Edwards, Department of Geography, Durham University, UK

Given the focus on responsive and incremental forms of adaptation in cities in the Global South, discussed above, it is perhaps not surprising to find that there are few examples of integrated climate-change-adaptation planning taking place in such cities. One well-known exception is the case of Durban, where efforts to foster evidence about climate vulnerability and climate adaptation have been in place for over a decade. In contrast to other cities, where mitigation has been the dominant concern, in Durban adaptation was an early priority for municipal climate-change responses, in the light of the potential co-benefits such a focus could bring in a context where issues of development and poverty were significant priorities (Roberts 2010). Following initial assessments of the potential impacts of climate change in Durban, work on adaptation began in earnest in 2006 with the development of a Headline Climate Change Adaptation Strategy (HCCAS). Its objectives were to identify the potential impacts and practical actions that could be taken to adapt to climate change in various sectors, including health, water and sanitation, waste, food security, planning, economic development and disaster risk reduction (Roberts 2010). The process identified the significant variation in interest and capacity in addressing climate change across these different areas and the ways in which traditional views of what each service was supposed to provide – e.g. short-term relief in the case of disaster risk reduction – hindered responses to climate change. Although the process was regarded as useful in bringing these tensions to light, 'it ultimately stimulated no new adaptation actions'

(Roberts 2010: 410). In seeking an alternative, the Environmental Planning and Climate Protection Department, which was leading climate responses in the city, sought to develop sector-specific adaptation plans as a practical means by which to overcome obstacles of scarce resources and a lack of political will for a more integrated approach. In essence, rather than building an integrated institutional structure and strategy through which to foster climate-change adaptation, the goal became that of building 'increased resilience one adaptation intervention at a time' (Roberts 2010: 401).

Although such an approach may appear to take one step back from those that advocate integrated approaches to adaptation planning as necessary for fully realizing its benefits, it does bring adaptation planning one step closer to implementation. In parallel with many national governments, urban adaptation planning remains largely confined to the gathering of knowledge and the development of strategies. Although such efforts signal the intention to act, they remain one step removed from the actual process of implementation (UN-Habitat 2011: 138). A recent survey in Canada, for example, found that, where municipalities have undertaken many of the measures and activities that precede policy implementation, only 4.5 per cent of those municipalities who are undertaking some form of adaptation have 'reached the point at which they completed "implementing projects and programs to reduce risk and increase resilience"' (Robinson and Gore 2011: 19). In this context, it is difficult to reach conclusions about what kinds of approach to adaptation planning – whether they are based on extensive knowledge of climate impacts or tuned to development priorities in a city, whether they seek to focus on specific 'at risk' sectors or adopt more integrated approaches – may be most effective in achieving adaptation. However, it appears clear that those strategies that have been developed to date have taken a more incremental, resilience-based approach to adaptation, and there is more limited evidence of adaptation being undertaken as a means through which to achieve transition and transformation (Birkmann *et al.* 2011).

Community-based adaptation: creating the potential for transition and transformation?

Engaging communities in adaptation is increasingly regarded as a critical part of any process that seeks to build resilience or to overcome existing forms of structural inequality. In those cities and urban areas where infrastructure deficits are significant, usually occupied by the poorest and most marginal communities (Chapter 2), the absence of governments or other external actors who can implement strategic and collective action means that adaptation is conducted in response to particular risks, as and when they occur. Furthermore, the needs of the urban poor, in the context of disasters or in terms of the provision of services for meeting everyday needs, are often neglected. In the Indian context, Aromar Revi (2008: 211) argues that, a 'chasm exists between the official urban "city building" development agenda and vulnerability reduction for those most at risk in these urban areas', resulting from the fact that the 'imperative of delivering adequate services (water, sanitation, solid waste, drainage, power) and equitable access to land and housing to the bulk of city residents is still a matter of contention'. In such contexts, fostering climate-change adaptation requires an understanding of the risks and vulnerabilities that communities experience, the provision of adequate services that meet basic needs, as well as specific strategies that seek to overcome structural inequalities in the city in order to include marginal communities in the processes of decision-making and implementation. Taking a community-based approach to adaptation is regarded as one effective means through which these three challenges – understanding climate risks, providing basic services and fostering participation in adaptation planning – can be met.

One critical role for community-based approaches to adaptation can be in providing knowledge about the nature of the risks and vulnerabilities that poor and marginalized communities face in different urban contexts. This knowledge, as discussed above, is often difficult to access, particularly in informal areas of cities where vulnerability may be most significant (Satterthwaite 2011). In the Philippines, the Homeless People's Federation is working with communities in twelve cities and ten municipalities to identify and profile vulnerability in informal settlements that may be most at risk, including those in locations prone to landslides and flooding (Satterthwaite 2011: 343). In Costa Rica, participatory research with local communities about the factors that lead to flood risks found that community members and those responsible for managing responses to emergencies had considerable knowledge about the causes of floods and of the particular factors that were shaping the incidence of flooding in their communities (López-Marrero and Tschakert 2011: 238). A community-based approach is also able to access knowledge about how individuals and households currently adapt to, or cope with, climate-related risks. In Dar es Salaam, researchers found a number of coping mechanisms in place for flood risk, including protecting assets and people through moving them to safer locations within the house or to higher ground (Dodman et al. 2011: 7). Although such responses are often small scale in nature and unable to address the root causes of vulnerability, they often form a vital strategy for responding to climate-related risks, particularly where local government is weak or ineffective. As such, supporting these forms of household response can be one means by which municipalities seek to work with existing, informal forms of adaptation and build capacity for resilience (UN-Habitat 2011: 131; see Box 6.3 and Figure 6.2).

In urban areas where infrastructure deficits are substantial, community-based organizations can also provide one means through which basic needs for food, water, sanitation and shelter are met. Although usually not undertaken

---

BOX 6.3

Informal adaptation in Accra

Climate change will affect Accra in Ghana in a range of different ways and exacerbate already existing socio-environmental hazards. With increased rainfall (both frequency and intensity) and other climate-change-related processes predicted along the Gulf of Guinea, a number of impacts are being experienced by the city. These include flooding and tropical storms, as well as an increased number of heatwaves and energy blackouts and increasing prevalence of disease such as malaria.

Many people in Accra live in poor neighbourhoods that lack the infrastructure or resources to adapt to these challenges, and this means that they are often the most vulnerable to these hazards and risks. Thus, communities in Accra are often forced to find ways to adapt to these new challenges, and informal adaptations to climate-change risks are a central part of how the city responds to these events. Communities and households in Accra develop their resilience to these risks through many different actions that show the resourcefulness and ingenuity of people who are often very poor. These informal actions, as opposed to municipal policies, include everything from putting stones on shack roofs to resist the high winds of storms, to residents clearing drains before the wet season, to local committees being set up to support community planning initiatives.

This informal adaptation by communities in Accra shows the importance of social networks and solidarity across the city in tackling the risks and hazards of climate change. With the ability of local and national governments to develop more formal adaptation actions curtailed by a lack of resources, these informal approaches provide an important level of support and protection across Accra and other African cities.

Jonathan Silver, Department of Geography,
Durham University, UK

Figure 6.2 Climate adaptation in Ga Mashie, Ghana
Source: Jonathan Silver

with the explicit aims of adapting to climate change, such initiatives can provide the basis upon which resilience can be built. For example, in Dar es Salaam, the Tanzanian Federation of the Urban Poor has developed small-scale savings schemes through which households can access capital to respond to climate-related shocks and stresses, provided the means through which communities have come together to seek alternative sites for housing, and undertaken small-scale initiatives to manage solid waste (UN-Habitat 2011: 134–5). This is just one example of the sort of work that is being undertaken by federations of slum and shack dwellers in different parts of the world, showing that, where 'there are representative community-based organizations, the possibilities of building resilience to climate change are much greater' (UN-Habitat 2011: 134). There remain, however, significant limits as to what community-based organizations can achieve, given the scale and scope of the actions required in many urban contexts. In some cities, partnerships between communities, community-based organizations and municipal authorities have been established as a means through which to develop approaches to adaptation that are more strategic and far-reaching in their aims and ambitions. One example is in Quito, where the municipality has 'provided funding to local NGOs to train indigenous farmers to improve the management of water resources in their urban agriculture practices and diversify as well as privilege native crops' (Anguelovski and Carmin 2011: 172). Another case is that of Ilo, where 'local government working together with community organizations has managed to develop safe and legal land sites that low-income households can afford' (Hardoy and Romero-Lankao 2011: 161). In the Philippines,

> community-based disaster preparedness is taking place as a way to better cope with typhoons and flooding events through awareness raising, early warning systems, and the creation of local institutions that provide residents with a safety net to cope with stresses stemming from natural hazards.
> (Anguelovski and Carmin 2011: 172)

Community-based adaptation, in principle, can provide new means through which local knowledge can be drawn upon in the development of adaptation strategies, and an approach through which the development needs of communities can be met while climate-change goals are realized. In practice, as with municipality-based adaptation plans, examples where such efforts are being realized in practice are less common. In many urban contexts, communities, and particularly those most vulnerable to the effects of climate change, remain marginalized from decision-making structures and are often neglected in the development visions of the city. At the same time, there remain critical limits as to what communities, in isolation, can achieve in response to the complex and uncertain challenges of climate change. In this context, there is a danger that the relevance of community-based adaptation will be 'both overstated and underplayed at the same time' (UN-Habitat 2011: 131).

Municipal climate-change adaptation in practice

> Most of the literature on climate change adaptation and cities is focusing on what should be done, not on what is being done (because too little is being done).
> (UN Habitat 2011: 145)

As set out above, owing to the dominance of mitigation concerns and the challenges of developing adaptation responses, strategies and measures to adapt to climate change are only now beginning to be developed at the urban level.

Furthermore, as we have seen above, even in those cities where adaptation planning is taking place, efforts to date have been put into assessing the potential risks from climate change to cities and particular urban economic and policy sectors and into developing strategies through which such responses could be delivered. In some cases, work has started on community-based climate-change adaptation, but here, too, there is considerable emphasis on gathering knowledge and, where appropriate, supporting communities in responding to climate-related risks as they unfold. It is for this reason, as UN-Habitat notes in its recent assessment of urban responses to climate change, that research on what is being done to adapt to climate change remains scarce.

Nonetheless, building on the concepts and ideas introduced in Chapters 4 and 5, it is possible to examine how municipal climate-change adaptation is taking place in practice by considering the different modes of governance that are being deployed and the sectors within which action has been forthcoming. Unlike mitigation, where municipalities have tended to adopt self-governing approaches, adaptation has been pursued through a combination of regulation, provision and enabling modes of governance. Examining three key urban sectors – urban development, built environment and infrastructure – provides some insight into the differences in the ways in which mitigation and adaptation have been pursued and their potential overlap. In addition, as discussed above, the close relationship between adaptation and responding to disaster risk means that this is another key sector in which urban adaptation is taking place. Across these different sectors, adaptation can involve 'building adaptive capacity' and 'implementing adaptation decisions, i.e. transforming that capacity into action' (Tompkins *et al.* 2010: 628). In what follows, these different modes of governance and types of action are considered across this range of urban sectors, before the chapter turns to identifying the opportunities and challenges that have shaped urban adaptation on the ground.

Responding to risk and disaster

Reducing exposure to risk and the impact of disasters has long been a focus of national and urban policy. The emergence of climate-change adaptation adds both complexity and challenge to this task. On the one hand, growing recognition of the impacts of climate change in terms of shifting the nature, magnitude and timing of long-standing risks, such as storms, drought or heat, may mean that existing approaches for managing risk are no longer adequate or, in some cases, may have to be fundamentally rethought. On the other hand, tried and tested means for responding to, coping with, and building resilience in response to, existing climate-related risks may provide a valuable source of knowledge and skills upon which climate-adaptation actions can build but which are not reducible simply to responding to particular climate impacts. Although disaster risk reduction and climate adaptation cannot therefore be regarded as one and the same, they share a common cause in 'reducing the impacts of extreme events and increasing urban resilience to disasters, particularly among vulnerable urban populations' (Solecki *et al.* 2011: 135). As a result, several initiatives that have been established to adapt to climate change have focused on enabling more effective responses to climate-related risks. These include approaches that seek to: reduce the exposure of cities to climate risks, discussed in more detail below in terms of urban development and urban infrastructure; provide early warnings of climate-related risks; and develop capacity to respond to and rebuild after disasters.

In terms of the development of warning systems, one sector in which several cities have developed responses concerns the development of plans through which to respond to heatwaves and to reduce the risks to human health. As discussed in Chapter 2, one potential impact of climate change is prolonged periods of extreme heat, exacerbated by the UHI effect, creating dangerous exposure to heat for urban residents. In Toronto, Chicago, Philadelphia, Paris and London, to name but a few, heat warning systems have been established in order to predict when urban weather conditions may be more likely to lead to increased levels of mortality and to trigger the use of health-protection measures (Box 6.4). These include the issuing of advice to individuals as to likely temperatures and cooling practices, the opening of 'cooling' centres in the city or, in the case of Paris's CHALEX plan, contacting elderly and vulnerable people directly, through networks of community health providers, such as health workers and pharmacists. However, research suggests that, 'heat emergency plans continue to neglect the poor and the socially isolated', and the intended practices of outreach, engagement and, where necessary, evacuation rarely happen in practice (Yardley *et al.* 2011: 676).

As discussed above, another approach through which disaster response and climate adaptation are being pursued in tandem is through the development of community-based responses to climate-related risks. This can involve community-based organizations, alone or working in partnership with municipal authorities, developing knowledge, resources and institutions through which households and neighbourhoods can respond to risk. It can also involve rather more top-down approaches, in which external agencies set up new institutions through which to engage communities in responding to risk. One example of this kind of initiative is the PROMISE-Bangladesh pilot project, funded by the US Agency for International Development (USAID),

BOX 6.4

Toronto's heat health alert system

How does the heat health alert system work?

From 15 May to 30 September, Toronto public health staff use the heat health alert system to determine whether a heat or extreme-heat alert should be declared. Forecast weather is compared with historical conditions that have led to increased rates of mortality in Toronto. A heat alert is issued when forecast weather conditions suggest that the likelihood of a high level of mortality is between 25 and 50 per cent greater than what would be expected on a typical day. An extreme-heat alert is issued when forecast weather conditions suggest that the likelihood of a high level of mortality is at least 50 per cent greater than what would be expected on a typical day.

What happens during a heat alert?

Once the Medical Officer of Health declares a heat alert, key response partners, community agencies and the public are notified. Hot-weather response activities focus on protecting vulnerable groups at increased risk for heat-related illness.

What happens during an extreme-heat alert?

In addition to the services provided during a heat alert, the city opens seven cooling centres when an extreme-heat alert is declared, some city pools may stay open longer, and health inspectors visit residential premises where heat risks may be high to ensure that the Hot Weather Protection Plan is being put into force.

Source: www.toronto.ca/health/heatalerts/alertsystem.htm

co-ordinated by the Asian Disaster Preparedness Centre and implemented by the Bangladesh Disaster Preparedness Centre in Chittagong. The project involves setting up 'ward disaster management committees, which included community members, school teachers, a local ward commissioner and local residents from higher-income groups', who are mandated to meet and determine actions to be undertaken locally under the terms of the national disaster-management plan (Ahammad 2011: 509–10). In addition to discussing potential responses, this pilot project has appointed a 'number of "change agents" . . . in 10 wards to carry out voluntary tasks with the ward committees regarding disaster risk reduction'; selected in consultation with local residents, these individuals have been 'trained to help the ward committees in pre-disaster preparedness and post-disaster response' (Ahammad 2011: 509–10). However, despite the potential of this role, doubts remain as to its effectiveness in a context where the needs of the urban poor are often marginalized, and where the capacity to co-ordinate with governmental and other bodies in order to achieve co-ordinated action to address long-standing challenges of urban development may be required (Ahammad 2011).

There are, therefore, very real challenges concerning municipal responses that can prepare adequately for risk and disaster. Doubts have also been raised about the extent to which the opportunities that disasters present for rebuilding and adapting cities have been utilized. Such opportunities arise, not from any linear, cause-and-effect relationship, but owing to the fact that such events lay bare the instabilities and inadequacies of current forms of development and create temporary vacuums of social and political authority (Pelling 2011a: 95). Rather than creating the means upon which to build more resilient alternatives, research suggests that, 'more commonly infrastructure is rapidly built back to pre-disaster conditions', because of the combined factors of a lack of awareness or knowledge of what these might be (Birkmann and von Teichman 2010: 176–7) and the vested interests in sustaining development as usual (Pelling 2011a).

Managing urban development

Where the risks of climate change are less immediate, other forms of urban response are taking place that seek to put urban development on a long-term trajectory for adapting to climate change. Two broad approaches can be identified here. In some cases, cities have sought to shape their physical development in such a manner that they can adapt to climate change. In other cases, the emphasis has been on the social and economic development of the city. Turning first to issues of physical development, some of the most important ways in which municipal authorities can shape current and future patterns of land use, and, hence, exposure to different kinds of climate-related risk, is through the use of the planning system (Davoudi *et al.* 2009; Measham *et al.* 2011; Wilson and Piper 2010). Although the particular competencies and resources that cities have in the planning domain vary significantly with national contexts, in most cases, municipal authorities have some responsibility for enabling or regulating the nature and location of (some forms of) urban development. In the context of developing countries, informal areas of housing and economic activity typically lie beyond the formal planning system. At the same time, planning frameworks that are in place, both in developing and developed countries, may not always be implemented in the way that they are intended, whether this is as a result of corruption or conflicts between planning authorities and local communities or businesses, or because of limited capacity to enact and enforce planning guidance and regulation. Although this means that the inclusion of climate adaptation within planning frameworks is no guarantee that adaptation is taking place in practice, the growing interest in integrating adaptation issues into urban planning is a sign that adaptation capacity is developing.

There are perhaps two sectors where the integration between planning and adaptation has been most advanced – coastal management and flood risk. The potential impacts of climate change in terms of sea-level rise and the increased intensity and frequency of storms have led to predictions that some cities will face challenges of adapting to coastal inundation and/or to river flooding (Chapter 2). In some cases, specific strategic-planning processes have been put into place in order to determine the levels of risk and the kinds of protective measure that should be adopted in order to reduce vulnerability. One example of this kind of approach is the Thames Estuary 2100 project (TE2100), established in 2002 by the UK's Environment Agency in order to develop a long-term tidal-flood-risk management plan for London and the Thames estuary (Environment Agency 2011). Under this project, which involved different public and private partners from the Greater London region, different 'decision pathways' were identified that detailed the 'thresholds at which various flood risk management measures fail to provide an acceptable level of protection' and the 'trigger points where a different approach to managing flood risk is required in response to a higher sea-level rise' (Nickson 2011: 6–7).

Other approaches have sought to integrate concerns for future climate impacts into existing planning processes for managing coastal regions. In the UK, municipal authorities have been involved in the creation of shoreline management plans, which 'provide a large scale assessment of the risks associated with coastal processes and present a long-term policy framework to reduce these risks to people and the developed, historic and natural environment in a sustainable manner' (Wilson and Piper 2010: 309). In Cape Town, the '*Coastal Development Guidelines* includes the so-called "blueline" below which new development should be prevented' (Cartwright 2008: 27), and, in Sydney, state-level sea-level-rise guidelines have been developed and subsidies have been offered to local councils in order that municipal planning processes can take climate impacts into account (Gurran *et al.* 2008; Measham *et al.* 2011: 905). Such measures are increasingly treading the fine line between the need to build defences to protect specific assets and populations and the realization that alternative forms of coastal management, such as realignment and retreat, may be both more cost effective and more sustainable in the long term (Wilson and Piper 2010: 307–14). Although awareness of the need for these kinds of change is growing, 'institutional legacies and ingrained socioeconomic interests' pose significant obstacles to this being realized in practice, and benefits that may accrue to one sector of society, such as a reduction in insurance losses, may not lead to overall social benefits, ensuring that the challenge of adapting to climate change in coastal regions remains contested and complex (Moser *et al.* 2008: 653).

Similar principles of the need to 'make space for water' have also begun to be advocated at the urban level in response to the risks of river flooding in the UK and the Netherlands (Wilson and Piper 2010: 287–99). This has been visible in the shift, over the past 15 years, away from a discourse on the need for 'large-scale engineering measures' to reduce flood risk towards a 'broader range of adaptation measures' (Harries and Penning-Rowsell 2011: 189). In the UK, municipal authorities are now charged with taking flood risk into account in the process of development, as well as with developing specific catchment flood-management and surface-water-management plans. In so doing, they must fully justify the use of particular sites for urban development and include flood-management measures in development proposals, such as the installation of sustainable urban-drainage systems. In London, the use of such measures has been mandated in a range of planning documents, including the overarching London Plan (City of London 2010: 16). Despite this integration, the uptake of this approach has proven to be relatively slow, in part as a result of the continued requirement from the insurance industry that adequate physical protection from flood risks is provided by the government before they will provide insurance, and because attempts to assert new approaches to managing flood risk have met with resistance from local communities who have been subject to flood risk (Harries and Penning-Rowsell 2011: 190).

The challenges of adapting to climate change through the planning process are thrown into even more stark relief in urban contexts where resources are limited. In Mexico, the Lerma river crosses some of the most densely populated regions of the country in the Toluca Valley (Eakin *et al.* 2010: 16). Traditionally regarded as a matter of managing the water resource for agricultural purposes, issues of urban development and the emergence of peri-urban areas are creating a complex landscape within which existing pressures for water use, coupled with changing climatic conditions, require new forms of management. Although, by law, municipal authorities are responsible for managing pollution entering the water course and preventing settlement in areas at risk of flooding, 'most municipalities are inexperienced in taking leadership to controlling growth and planning for land-use change' and 'often lack the equipment and resources . . . needed to implement the recommended actions' (Eakin *et al.* 2010: 18). As a result, responsive measures are used to address flood risk, which not only do not address underlying problems but may also increase risk 'by providing temporary protection for areas that should not be settled in the first place' (Eakin *et al.* 2010: 21). In other contexts, municipal authorities may actively seek to protect some areas of the city at the expense of others. Chaterjee (2010: 344) suggests that, in Mumbai, public funds have been directed at protecting the formal economic sector and global businesses from flood risk, rather than affording additional protection to the residents of informal settlements in the city, who instead bear the consequences of such interventions by being uprooted or displaced. Such issues of who should benefit, and who may lose, from adaptation are critical. As Huq *et al.* have argued, the

> kinds of changes needed in urban planning and governance to 'climate proof' cities are often supportive of development goals. But . . . they could also do the opposite – as plans and investments to cope with storms and sea-level rise forcibly clear the settlements that are currently on floodplains, or the informal settlements that are close to the coast.
>
> (2007: 14)

Given the significant institutional, political and social challenges of integrating adaptation into existing frameworks for urban planning and economic development, approaches that seek to put community needs first may offer one way of taking adaptation forward. In Bangladesh, the Community Based Adaptation to Climate Change through Coastal Afforestation (CBACC-CA) project has been developed in four coastal communities, including Chittagong. The aim of the project is to create a buffer to climatic hazards through the development of mangroves, plantations, dykes and embankments, together with early-warning systems and disaster-preparedness systems (Rawlani and Sovacool 2011: 859).

In addition, it seeks to build social capacity through promoting forestry, fishing and farming, as means of developing new livelihoods and adapting to climate change at the same time, 'by integrating aquaculture and food production within reforested and afforested plantations' (Rawlani and Sovacool 2011: 859; see also Kiithia 2011: 178). In Durban, approaches have also been developed that seek to integrate responding to climate change with the development priorities of the city, and particularly the need to ensure food and water security under changing climatic conditions (Box 6.5). Despite the promise of such approaches, the challenges of community-based approaches discussed above, including their often incremental and small-scale nature and their inability to address more structural processes that shape development trajectories, must not be overlooked (Rawlani and Sovacool 2011: 860). All too often, processes of urban development continue without adequate attention being paid to the need to take climate adaptation into account, and they do so while continuing to put the most vulnerable urban residents at greatest risk (Manuel-Navarrete *et al.* 2011).

Designing and using the built environment

At the scale of individual buildings, various forms of adaptation have also been undertaken. Here, adaptation measures tend to be concerned with improving the resilience of the building to different forms of risk – including flooding, storm damage and heat – or focused on using the building envelope as a means through which to secure resources, such as water or cool air, that may be limited under changing climatic conditions. Such measures can be formally mandated, enforced through the regulatory powers of municipal authorities, or promoted by municipalities through forms of engagement and education. At the same time, urban residents may employ a range of such measures as a means of coping with particular risks or in order to become more self-sufficient under conditions of resource scarcity.

Formal measures that have sought to improve the resilience of buildings to climate-related risks include the inclusion of particular design guidelines or standards within building codes and planning documents. One such example in the UK context is the inclusion of recommendations for building designed to recover quickly from the impacts of flooding through ensuring that essential services (power, water, sanitation) experience minimal disruption (e.g. by placing power sockets above likely flooding levels) (Department for Communities and Local Government 2007). Where vulnerability to heat is of concern, efforts are being made to design different forms of 'cool roofs' – roofs that deflect incoming solar radiation and/or provide high levels of

BOX 6.5

Community-based climate-resilient development in Durban

In order to foster adaptive capacity, pilot-scale community-based adaptation projects have been initiated in two 'poor, high risk, low-income communities, namely Ntuzuma, representing the more urban sectors of the city, and Ntshongweni, representing the more rural areas' in Durban. These pilots have involved three types of intervention:

- community-based adaptation planning: in order to develop 'a more detailed understanding of community level risk' and as the basis for establishing 'community level action plans' a variety of approaches have been employed, including risk and vulnerability mapping, awareness raising, and assessment of the sustainability of different adaptation options;
- food security: the 'productivity of dry land maize (a key subsistence crop) will drop to almost zero under projected climate change conditions', and there is a need to identify alternative crops. A process has been undertaken that seeks to assess food security, to undertake field trails of alternative crops, and the 'development of a food security action plan';
- water security: under current climate-change scenarios, water availability is likely to be reduced under climate change. In this context, there has been a growing interest in 'identifying workable and sustainable water-harvesting technologies', and this has involved examining technologies that can improve food security in poor communities through field testing and evaluation.

In addition, through Durban's commitment to being carbon neutral during their hosting of the 2010 Football World Cup, opportunities were identified to undertake 'reforestation projects that will not only result in carbon sequestration (through tree planting) but will also alleviate poverty and address issues of environmental degradation and catchment management.' These projects have involved the planting of over 80,000 trees and have generated benefits for local communities through the 500 'treepreneurs' that have been established to supply the tree seedlings to the project. The 'treepreneurs' trade their tree seedlings for food, school fees and other basic goods at 'tree stores', thereby providing income generation opportunities for impoverished communities living far from zones of economic opportunity.

Source: Roberts 2010: 408–10

insulation to keep buildings cool and reduce the effect of the UHI. In California, provision has been made under the Building Energy Efficiency Standards for buildings to incorporate white or reflective roofing materials as a means of cooling buildings without using air conditioning (The California Energy Commission 2012). The Global Cool Cities Alliance, initiated by researchers at the Universities of Berkley and Concordia, seeks to recruit 100 cities globally to undertake commitments to the widespread roll out of cool-roof technologies (Global Cool Cities Alliance 2012). Cities such as New York and Philadelphia have established programmes and initiatives to implement cool roofs on public buildings, engaging a self-governing approach, and to enable the uptake of cool roofs by individuals and businesses (see the city case study on Philadelphia above). One specific form of cool-roof technology that is increasingly being integrated into urban planning and building codes is that of green roofs. A green roof is 'a layered system comprising a waterproofing membrane, growing medium and the vegetation layer itself' that can reduce storm-water run-off, cool buildings, enhance biodiversity and extend the lifetime of the roof itself (Castleton *et al.* 2010: 1583). The multiple benefits of green roofs have meant that municipal authorities are increasingly seeking to promote their implementation. For example, in Chicago, the Department of Environment established a Green Roof Grants Program that funded over seventy green-roof projects between 2005 and 2007.

In addition to providing the means through which to increase resilience to risks of flooding and heat, the built environment can also provide a means by which adaptation to water stress or shortage can take place. In urban areas where water supplies are currently under stress because of climatic conditions, or where water services are fragmented or non-existent, new forms of water supply and sanitation are emerging. One such development is the recycling of grey water, using water that has previously been used for one household purpose (such as bathing or washing up) for other purposes, including gardening, car washing or toilet flushing. In some cases, such systems are informally developed, particularly during times of water stress, when the use of fresh water for some such purposes is banned. In other buildings, new systems for recycling grey water are included in the design and construction, such as the example of Melbourne's Council House 2 (CH2), which includes a grey-water and black-water treatment plant and 'mines' a nearby sewer for additional water resources (City of Melbourne 2012). Another means through which buildings are being used to capture water is through rainwater harvesting. Box 6.6 and Figure 6.3 illustrate the development of so-called 'rainwater-harvesting' schemes in Mumbai, where, in common with other Indian cities, rainwater-harvesting schemes are becoming an increasingly common means by which to meet the growing demands for water in a rapidly urbanizing context.

BOX 6.6

Rainwater harvesting in Mumbai

The ancient technology of rainwater harvesting is undergoing a revival and reinvention in urban India owing to water shortages. The simplest form of rainwater harvesting is to guide run-off into a tank. A natural slope can be used to guide storm-water into a channel, and water can also be collected from the roof. In some historic buildings, rainwater is collected on the roofs and delivered to underground tanks via ornate spouts. Storing water in this way is problematic in Mumbai, as the monsoonal rain pattern necessitates a large-capacity tank, and holding still water can lead to mosquitoes breeding.

Rainwater is used to recharge wells in many systems. Rainwater can be collected from the roof of a building and guided through filters before being used to recharge a bore well. Modern pre-cast wells and recharge pits draw on designs of stepwells in rural Rajasthan, re-imagined to fit urban requirements. This method uses up very little space, and the wells are usually placed underneath parking bays. The domestic water supply is differentiated in Mumbai, so that the harvested rainwater can be used for specific applications, and bore-well water is mainly used outdoors, for example for washing cars, and for toilet flushing. This system is the type most commonly retrofitted into middle-class apartment blocks to increase water security and reduce costs, as well as to address environmental concerns. Another option is to direct the rainwater to soakaways that replenish the groundwater. These systems collect water from gardens and direct it to special areas where it soaks into the earth, rather than running out to sea. In large apartment buildings, a combination of methods might be used, including percolation into lawns and combination with grey water.

Catherine Button, Department of Geography,
Durham University, UK

Figure 6.3 Concrete recharge pits and bore wells in Mumbai, designed using the principles of Rajasthan stepwells
Source: Catherine Button

Beyond these formal initiatives, the built environment is also a domain in which informal practices of coping and adaptation take place. In Mumbai, in the absence of effective urban development strategies for reducing risk in informal parts of the city, individual households undertake various structural improvements to reduce the risks that they may encounter. Research found that 'after the 2005 flood event approximately 53 per cent of households surveyed raised their foundation before monsoons in 2006 to ensure that the event of previous year is not repeated', and that other structural measures, such as creating elevated areas for storing valuable assets, were also undertaken (Chaterjee 2010: 345). Undertaking such adjustments has a cost, and not all households were equally able to undertake such measures, so that, in any one street or neighbourhood, differences in adaptive capacity lead to different levels of vulnerability (Chaterjee 2010: 346). Other forms of informal adaptation are also found in a range of urban contexts. Taking the example of heat risk, these range from physical or technical measures, such as shading and vegetation to cool particular spaces or the use of cooling technologies, to other forms of behavioural change, including dressing in lighter clothes or changing patterns of eating, working and sleeping. Beyond the notion of adaptive capacity, which focuses on the knowledge, skills and resources that individuals have and their economic and social context, factors that shape the cultural and social practices of adaptation are less well understood, but may relate to institutionalized and culturally specific ideas of what constitutes 'normal' behaviour (Adger *et al.* 2009). In this sense, adapting to climate change may require challenging and changing current conventions and routines.

Reconfiguring urban-infrastructure networks

The impacts of climate change will have particular significance for urban-infrastructure networks – those systems that provide water, sanitation, energy, communications and transport services. In the cases of water, sanitation and energy systems, the provision of urban services can be directly related to the availability of water and its availability for direct use or for use in the production of electricity or hot water. In all cases, the impacts of climate change may lead either to gradual encroachment on networks (through, for example, saltwater intrusion in groundwater supplies in coastal cities), or to specific impacts associated with climate-related events, including tropical storms, droughts, flooding and heatwaves. Efforts to adapt urban-infrastructure networks to climate change have involved the development of adaptation capacity, through assessments of potential risks and new forms of decision-making that seek to take climate change into account, as well as specific efforts to enhance the physical and social resilience of specific networks.

As discussed in Chapter 2, New York is one city where the potential impacts of climate change on urban-infrastructure networks have been extensively examined (Box 6.7). The New York City Department of Environmental Protection (NYCDEP), the administrative body responsible for water supply and sewer- and wastewater-treatment systems in New York, established a Climate Change Task Force in 2004 as a partnership between scientists and policymakers. Through this process, a 'framework for analysing climate

> **BOX 6.7**
>
> **Integrating climate-change adaptation into infrastructure planning in New York**
>
> The results of the NPCC process show that there are numerous ways that climate-change adaptation can also be effectively incorporated into the current management of the city's critical infrastructure:
>
> - Existing risk- and hazard-management strategies can be adjusted to meet the challenges of our changing climate – today and in the future.
> - Design standards can be recalibrated to include climate-change projections, so that long-lasting infrastructure will be prepared to withstand future threats.
> - The legal framework governing the design and operation of infrastructure can be expanded to include the impacts of climate change.
> - The insurance industry and other risk-burden-sharing mechanisms (e.g., co-operatives) can contribute to adaptation through products that respond to long-term risks, as well as by sharing expertise on risk in climate-change discussions with a wide range of stakeholders.
> - Within and across agencies and organizations that manage infrastructure, adaptation strategies can draw from a broad range of responses, including adjustments in operations and management, capital investments in infrastructure, and development of policies that promote flexibility.
>
> Source: Rosenzweig and Solecki 2010

change has been created, including a 9-step Adaptation Assessment procedure', under which 'potential climate change adaptations are divided into management, infrastructure, and policy categories, and are assessed by their relevance in terms of climate change time-frame (immediate, medium, and long term), the capital cycle, costs, and other impacts' (Rosenzweig *et al.* 2007: 1400), in order that suitable options can be chosen and implemented. This has led to a study of the impacts of sea-level rise on sewer and wastewater treatment, and to new approaches to modelling water availability in the New York catchment area that take account of the effects of climate change (Rosenzweig *et al.* 2007: 1407). In Canada, research has found that measures are already being taken in anticipation of reduce levels of water availability in Ontario in the cities of Waterloo and Guelph, and these include demand-management measures such as 'voluntary and mandatory outdoor water use restrictions, water rationing, public education, water pricing, and installation of water-saving equipment' (de Loë *et al.* 2001: 236). Such measures involve the realignment of different institutions and the integration of new forms of knowledge into decision-making. While these initiatives may focus on either physical or social interventions, the very process of making decisions that take adaptation into account requires the development of adaptive capacity at the institutional level. In the case of Ontario, as elsewhere, this capacity is frequently limited by the absence of necessary legal or institutional powers (de Loë *et al.* 2001: 236).

In urban areas where such services are less well developed or absent, it is a deficit of well-functioning infrastructure that poses the most significant challenge to adaptation, as discussed above. In Mumbai, Chaterjee (2010: 346) found that residents engaged in various forms of pre-emptive coping measures, including efforts led by local groups collectively to clean, widen and cover drains before the annual monsoon. In Dar es Salaam, a new scheme whereby the municipality collects solid waste in unplanned settlements for a nominal payment by residents is also helping to address the challenges of maintaining adequate forms of sanitation and drainage in these areas of the city (Dodman *et al.* 2011: 9). However, although such initiatives are to be welcomed in terms of reducing immediate risks, as discussed above such forms of coping and incremental resilience cannot address wider challenges of vulnerability. Such settlements are often located in environmentally marginal locations that are particularly vulnerable to risk, so that permanently addressing the climate-related hazards to which communities may be exposed is impossible. The challenging question in such cases is whether some form of relocation, with all of the disruption to social and economic lives that it might entail, is therefore necessary in order to respond to climate change. Much, of course, depends on how such resettlement might be carried out – forcibly evicting vulnerable people who have nowhere to relocate to could not be considered as a sustainable form of adaptation. However, achieving the forms of transition and transformation that Pelling (2011a) suggests are required in order to be able to develop *with* climate change will require that adequate shelter and services for the urban poor are given a much higher priority.

In addition to addressing the challenges of existing infrastructure, one response to climate change at the urban level has been the development of new forms of green (vegetation) and blue (water) infrastructure system. This can range from the scale of individual buildings, where, as discussed above, green roofs are increasingly popular, to urban-wide strategies for developing the amount of open space, recreational water use and trees in cities. In Chicago, the adaptation strategy includes a target of increasing the urban tree canopy by 20 per cent by 2020 in order to reduce the impact of the UHI effect, while also sequestering carbon dioxide from the atmosphere. In other cities in the US, including New York and Los Angeles, schemes to plant 1 million urban trees have also been established. Such approaches are also evident in European cities, such as Stuttgart and Freiburg in Germany, where principles for developing and including open space through which to develop green infrastructure for cooling the city and providing the means for dealing with excess storm-water, have been included within municipal plans (Carter 2011: 194). In Latin America, schemes to plant urban trees and protect peri-urban areas of forest and vegetation have also been established, although these are most often framed in terms of their mitigation potential (Chapter 5).

Across the areas of disaster response, urban development, the built environment and urban-infrastructure systems, it is evident that some forms of adaptation are emerging. In some urban places, particularly those characterized by infrastructure deficits and ongoing challenges of economic, political and social exclusion, responses are more akin to coping. Although individual initiatives that seek to build resilience and engage communities in a more transformative manner are visible, they remain the exception. In more economically developed urban areas, there is evidence that municipal authorities and their partners are engaging in more sustained approaches to building adaptive capacity and to taking this into account in urban planning and regulation, and in the ways in which they provide services and seek to enable others to respond to climate change. Given that there is limited evidence of action on adaptation at the municipal level, it is currently not possible to determine which of these different modes of governance – regulation, provision or enabling – is becoming dominant. However, it is clear that, as with mitigation, adapting to climate change in the city is being driven by a range of different factors and encountering some significant challenges.

Drivers and barriers for urban climate-change adaptation

As Chapter 4 sets out, the drivers and challenges of responding to climate change could be considered as falling into three broad areas: institutional, political and sociotechnical. Underpinning many of the concerns about the limited extent to which adaptation has been, and can be, developed at the urban scale is the knowledge that, 'local authorities lack fiscal resources, decision making power and other components of the institutional capacity to address multiple and reinforcing development problems as well as the underlying or root causes of vulnerability' (Hardoy and Romero-Lankao 2011: 161), even before climate change is considered. Although institutional capacity, in its broadest sense, is important for addressing this challenge, political considerations – how and why vulnerability is conditioned, and the urban-adaptation agenda framed and enacted in particular ways – as well as the ways in which existing urban sociotechnical systems serve to enable and constrain livelihoods and responses, also need to be taken into account (Table 6.2).

At root, most of the *institutional* drivers and barriers that have been identified as critical in shaping urban adaptation are related to the limited resources that exist in most municipalities to address even the basic needs of urban populations, let alone to engage with an issue that is regarded as of less immediate concern or local relevance (Roberts 2010). In Indian cities, Sharmar and Tomar (2010: 461–2) find, not only that local governments 'struggle with other development pressures and hardly find global concerns such as climate change of any interest', but also that there is a 'gross lack of capacity within local governments to facilitate the institutionalization of issues such as climate change'. In this context, where climate change appears to be a distant issue of little immediate concern and where capacity is severely limited, it is perhaps unsurprising that climate adaptation has yet to become mainstream at the urban level. However, the extent of local knowledge about the potential impacts and vulnerabilities of cities in relation to climate change can be both a driver and barrier to adaptation. In those cities where formal assessment processes have taken place, including, for example, New York, London and Durban, adaptive capacity has been built, and new adaptation measures are often in place. However, in the majority of cities, uncertainty about the timing and impact of climate change, and the potential costs of adaptation are regarded as a barrier to action (Carter 2011). In particular, where resources are scarce, the challenges of calculating the costs of action, and of inaction, as well as the timescales over which benefits will be realized, mean that making the case for investing in adaptation measures is often very challenging. Under such conditions, rather than seeking to realize holistic adaptation strategies, alternative approaches that focus on, for example, revising building codes, land-use management, changing infrastructure standards etc. may provide a means through which greater resilience can be built without significant up-front costs (Satterthwaite 2008b).

Table 6.2 Drivers and barriers for urban climate-change adaptation

	Urban development	Built environment	Urban-infrastructure systems
Institutional drivers	Ownership of land/stake in development project. Supportive national and regional planning frameworks. Partnerships with proactive private-sector urban-development organizations. Access to sufficient capital to address additional upfront costs of adaptation	Own/operate housing and commercial stock. Supportive national and regional policy goals. Partnerships with private and civil-society organizations. Exchange of knowledge with other cities. Availability of external funding for initiatives	Own/operate infrastructure systems. Supportive national and regional policy goals. Partnerships with private and civil-society organizations. Exchange of knowledge with other cities. Access to capital for undertaking large-scale climate-protection measures
Institutional challenges	Limited capacity and resources – knowledge, people, finance. Uncertainty regarding the extent, location and timing of climate impacts and persistent perception that climate change is a distant issue. Lack of knowledge about nature and extent of informal urban development. Absence of adaptation concerns within local, regional or national planning frameworks. Implementing and enforcing planning regulations. Lack of policy co-ordination and conflicting policy goals. Mismatch between urban jurisdiction and urban-growth pressures. Lack of effective or accountable municipal government, particularly for residents in informal urban settlements	Limited capacity and resources – knowledge, people, finance. Uncertainty regarding the extent, location and timing of climate impacts and persistent perception that climate change is a distant issue. Lack of knowledge about existing vulnerabilities. Absence of adaptation considerations within existing regulation. Problems with the implementation and enforcement of regulations. Lack of policy co-ordination and conflicting policy goals. Lack of effective or accountable municipal government, particularly for residents in informal urban settlements	Limited capacity and resources – knowledge, people, finance. Uncertainty regarding the extent, location and timing of climate impacts and persistent perception that climate change is a distant issue. Lack of knowledge about the ways in which impacts in one system may affect others. Lack of knowledge about existing vulnerabilities. Lack of municipal competencies in key infrastructure sectors. Absence of adaptation considerations in the institutionalized rules and mechanisms shaping investment decisions and operating practices for infrastructure systems. Lack of policy co-ordination and conflicting policy goals. Mismatch between urban jurisdiction and factors shaping development of urban-infrastructure systems. Lack of effective

continued ...

Table 6.2 ... continued

	Urban development	Built environment	Urban-infrastructure systems
			or accountable municipal government, particularly for residents in informal urban settlements
Political drivers	Leadership – flagship projects, advancing city's profile internationally. Co-benefits for urban-development agendas – e.g. reducing disaster risk, enhancing urban green space, climate mitigation. Windows of opportunity for large-scale redevelopment projects. Opportunities for municipalities to engage communities and stakeholders in the design and implementation of adaptation strategies/measures. Presence of intermediary or community-based organizations with sufficient capacity to organize and address vulnerability	Leadership – opportunities to address concerns of key urban constituencies through e.g. reducing health vulnerabilities, improving housing conditions. Co-benefits – e.g. reducing disaster risk, addressing poverty, enhancing building fabric, climate mitigation. Windows of opportunity for new buildings/ interventions in existing buildings. Opportunities for municipalities to engage communities and stakeholders in the design and implementation of adaptation strategies/measures. Presence of intermediary or community-based organizations with sufficient capacity to organize and address vulnerability	Leadership – flagship projects that promote e.g. energy independence, modernization of transport systems. Co-benefits – e.g. reducing disaster risk, security, meeting basic needs, climate mitigation. Windows of opportunity for development of new infrastructure systems. Opportunities for municipalities to engage communities and stakeholders in the design and implementation of adaptation strategies/measures. Presence of intermediary or community-based organizations with sufficient capacity to organize and address vulnerability
Political challenges	Absence of leadership or political will. Conflicts with agendas for urban development and growth. Deliberate neglect and marginalization of the interests and agendas of the urban poor. Competition for funding and attention with what are regarded as more urgent and immediate	Absence of leadership or political will. Conflicts with other agendas for urban (re)development and retrofitting. Deliberate neglect and marginalization of the interests and agendas of the urban poor. Competition for funding and attention with what are regarded as more urgent and immediate	Absence of leadership or political will. Conflicts with other agendas to provide services and secure resources. Deliberate neglect and marginalization of the interests and agendas of the urban poor. Competition for funding and attention with what are regarded as more urgent and immediate development needs. Lack of access to decision-making arenas for

	development needs. Lack of access to decision-making arenas for poor and marginalized groups. Mismatch between long-term process of urban adaptation and short-term political cycles		poor and marginalized groups. Mismatch between long-term process of developing new infrastructure projects and short-term political cycles
Socio-technical drivers	Existing urban morphology that is conducive to adaptation; adaptation measures that can be readily incorporated into urban landscape. Emergence of niches/experiments for alternative technologies and social organization. New sociocultural expectations and practices for urban living that favour sustainability	Built environments that are conducive to forms of intervention for adaptation; adaptation measures that can be readily incorporated within existing systems, structures and practices. Emergence of niches/experiments for alternative technologies and social organization. New sociocultural expectations and practices for buildings that favour sustainability	Well-functioning and well-maintained infrastructure networks; adaptation measures that can be readily incorporated within existing systems, structures and practices. Emergence of niches/experiments for alternative technologies and social organization. New sociocultural expectations and practices for energy, water and waste services that favour sustainability
Socio-technical challenges	Urban morphology based on exposure to climate-related risks and sea-level rise. Urban-development pressures that lead to informal settlements in marginal areas of cities that are exposed to climatic risks. Continuation of urban-development and -planning practices based on historical climatic conditions. Mal-adaptations that constrain adaptive capacity, e.g. continued development in floodplains. Adaptation measures that require significant reconfiguration of urban landscapes and encounter contestation and conflict	Historical legacies that make retrofitting and redevelopment of buildings costly or challenging to existing urban aesthetics. Lack of adequate housing and shelter, particularly for vulnerable communities. Adoption of historically/culturally based responses to climate-related risks that may no longer be appropriate. Mal-adaptations that constrain adaptive capacity, e.g. building in future demand for water in drought-prone areas. Adaptation measures that are incompatible with existing built form, systems and practices, and that encounter contestation and conflict	Infrastructure deficit and failure to meet basic needs. Rigid infrastructure networks and institutional cultures that favour incumbent technologies and prevent change. Continuation of organizational and behavioural cultures in the development and use of infrastructure networks based on past climatic conditions. Mal-adaptations that constrain adaptive capacity, e.g. location of critical infrastructure projects. Adaptation measures that are incompatible with existing infrastructure systems and practices, and that encounter contestation and conflict

As with issues of mitigation, another set of institutional drivers and barriers relates to the multilevel governance of adaptation. As a cross-cutting issue, which affects sectors as diverse as health and energy, transport and recreation, co-ordinating an effective adaptation response within a municipal authority and across relevant partner agencies can be difficult. In Durban, Debra Roberts found that the challenges of a lack of skills, finance and knowledge for climate-change adaptation were 'exacerbated by the implicit (and often explicit) assumption that environmentally related issues such as climate change will be dealt with by the EPCPD, so there is no need to engage with them in any depth' (Roberts 2010: 401). Furthermore, adaptation challenges often cross multiple jurisdictions. This can take place vertically, for example in the case of investment in the protection of critical infrastructure that may require national, regional and local government finance and approval, or horizontally, for example between local authorities over the course of a river. Allocating responsibilities and co-ordinating action for adaptation between the different government and non-governmental bodies involved are also significant challenges. In Europe, for example, Carter (2011) suggests that the presence of an overarching strategy for adaptation at the level of the European Union has not been matched by the development of national and local plans. At the same time, where adaptation is emerging, in coastal or spatial strategies, bottom-up approaches are, to a large degree, shaped by national and European policies, which both provide generic support for the principle of adaptation and serve to limit the scope for specific actions (Carter 2011). Further complexity is also added by the *lack* of effective or accountable governance structures for many of the most vulnerable populations in cities, where their status as occupants of informal settlements means that they are overlooked by urban authorities.

The *political* drivers and barriers to urban climate-change adaptation include challenges of leadership, of making adaptation a relevant and local issue, and of engaging with the needs of vulnerable urban communities. Unlike mitigation, where politicians and policy entrepreneurs have been able to demonstrate innovation, and leadership has provided a critical driver for urban responses, in the adaptation arena there is an absence of such opportunities. The sorts of action that may lead to reduced vulnerability, such as the clearing of drains or addressing everyday challenges of health and development, do not lend themselves to newspaper headlines and are not the stuff of city-based competitions. In short, beyond large-scale infrastructure projects, there is limited political capital to be made from many adaptation measures, and they have, to date, not attracted the same kudos for promoting cities internationally as have those efforts to develop low-carbon cities. What appears to be required is a different form of leadership, one that focuses on enhancing participation and inclusion and that attends to the needs of marginalized and vulnerable communities, particularly the urban poor. Unfortunately, research suggests that, in many of those places where such leadership is urgently required, it is found wanting, as municipal governments overlook or explicitly disregard the needs of the most vulnerable, particularly those who inhabit informal areas (Satterthwaite 2011). Research on local-scale adaptation has found that, 'institutional barriers to spaces of political power and governance can impose severe restrictions on adaptive capacity for certain fragments of the community' (Jones and Boyd 2011: 1271).

Particularly, but not exclusively, in relation to places in which such processes of marginalization are persistent, the presence of active 'intermediary' organizations, such as NGOs or development charities, and community-based organizations has been critical in creating opportunities for engagement and in driving forward alternative forms of adaptation (Dodman *et al.* 2011; Pelling 2011b; Satterthwaite 2011; UN-Habitat 2011). However, as is evident in the examples discussed above, such organizations often lack the capacity and resources to go much beyond enhancing existing coping mechanisms or undertaking small-scale actions. As Hardoy and Romero-Lankao (2011: 161) suggest, although 'existing experiences illustrate that the co-ordinated work between government and civil society is more likely to bring about effective responses', such efforts are far from common, because 'the channels and vehicles of participation are rarely there to support this kind of coordinated work'. Moving beyond current contexts where the concerns of vulnerable populations are excluded requires developing frameworks and governance structures that link adaptation with broader agendas of urban renewal and that address 'existing asymmetries and structural vulnerabilities', a challenge that will require 'strong interventions in real estate and housing markets and public service delivery, and a supportive policy and institutional environment at state level' (Revi 2008: 222).

Whether adaptation is municipally driven or led by intermediary and community-based organizations, it is 'important to acknowledge that pre-emptive adaptation to potential future climate change impacts is often not the driving force' behind initiatives and projects (Carter 2011: 195). Rather, other agendas, including the need to develop urban green space, address development needs, respond to disaster risk, replace ageing infrastructure or address current issues of resource costs or security, may provide the framework and motivation for actions that have the co-benefit of reducing vulnerability and enhancing adaptive capacity. Unlike the mitigation arena, where co-benefits are usually sought for policy agendas driven by explicit climate-change concerns, in the adaptation arena, it appears that adaptation itself emerges as a co-benefit of other policies and projects. Although this may provide one means of achieving adaptation with low political costs, it also means that urban development pathways that exacerbate vulnerability may continue unchecked. Such challenges may be exacerbated where there are conflicts over how problems are framed and interpreted, over where responsibilities lie, and where vulnerable groups are excluded from decision-making (Eakin *et al.* 2010: 16). Manuel-Navarrete *et al.* (2011) suggest that, in the Cancún coastal region of Mexico, an economic-growth pathway focused on mass tourism has exposed ever-greater numbers of people to the climate-related risks of hurricanes, although such vulnerabilities have been secured through 'top-down coping strategies', including a high-profile 'civil protection system' designed both to address disasters and 'to preserve the image of a "safe destination" amongst international tourists and operators' (Manuel-Navarrete *et al.* 2011: 257) and access to external sources of funding and insurance that enable restoration and repair to take place in the aftermath of hurricanes. The dominance of political economic interests in the mass-tourism sector of the economy, and its reliance on 'maintaining a good image for the touristic destination leads to prioritizing effective command-and-control strategies that avoid casualties and generate a broad sense of safety without addressing the roots of differential vulnerabilities' (Manuel-Navarrete *et al.* 2011: 257). In essence, particular sets of political and economic interests serve to promote the resilience of existing forms of economic development, without considering more transformative forms of adaptation that would also address existing patterns of vulnerability and exclusion.

Such political challenges to adaptation take place within the *sociotechnical* contexts in which urban systems and practices are configured (Chapter 4). Different kinds of urban morphology, whether cities are low lying, situated in delta regions, dependent on particularly vulnerable forms of energy or water supply, for example, all serve to structure the broad context within which vulnerability is created and adaptation takes place. Moreover, as discussed in detail above, the extent to which urban-infrastructure networks are adequate and functioning, or whether significant infrastructure deficits exist has been shown to be a critical factor shaping levels of vulnerability and adaptive capacity (Satterthwaite *et al.* 2008b). Such existing systems, as well as the continuation of historical and cultural practices of urban development, planning, organizational practices, customary building methods, livelihoods and forms of production and consumption that are structured around existing climatic conditions, serve to continue existing forms of vulnerability and to create forms of 'mal-adaptation', 'climate adaptation constraining decisions or actions' (Carter 2011: 195). Where such systems, structures and practice are rigidly maintained and inflexible, the potential for interventions that seek to reconfigure, for example, building standards, planning zones or water use will be limited. Where adaptation measures can, for example, serve 'everyday as well as disaster risk functions', such as 'stairways and bridges that enabled market access and greater social interaction ... in addition to providing emergency access', they may be more able to be integrated with existing sociotechnical networks and serve to enhance resilience (Pelling 2011b: 399). Small-scale niches or experiments can provide the means for opening up such systems to alternative possibilities, forging a more transitional or transformative approach to adaptation, but the extent to which they can create widespread change has yet to be evaluated (see Chapter 7).

Conclusions

In comparison with mitigation, explicit urban responses to climate-change adaptation have, to date, been undertaken by a much smaller number of cities and remain relatively underdeveloped. The international focus on issues of mitigation, together with the diffuse and complex nature of adaptation, appears to have constrained urban responses beyond a group of pioneering cities. However, there is evidence that interest in adaptation as a specific policy domain is growing, with the increasing engagement of transnational municipal networks and donor agencies with the adaptation agenda. At the same time, there is a growing recognition that many of the policies and measures undertaken by municipalities towards other ends, including addressing disaster risk and development needs, need to be developed with climate adaptation in mind. Although the dominant model of urban adaptation remains focused on the creation of expert assessments of the impacts of climate change and urban-planning frameworks through which these can be taken into account in particular cities, there is evidence that alternative, community-based models of adaptation are also emerging that focus on understanding the social and economic dimensions of vulnerability in the city and that seek to address this through developing adaptive capacity at the grass-roots level. The evidence and examples presented in this chapter suggest that it may be where these different approaches can be brought together that the most significant progress in undertaking urban climate-change adaptation is taking place.

Despite some evidence of progress and innovation, however, there remain significant challenges to achieving adaptation in practice. Although the blurred boundaries between climate-change adaptation and 'good' forms of development and governance can be advantageous, in terms of offering a means by which to bring climate-change considerations into mainstream urban agendas, whether that be for providing basic services or protecting critical infrastructures, it can also serve to diffuse climate-change concerns among other competing priorities and obscure the need to carry out urban development differently in order either to attend to future risks or radically to transform existing forms of vulnerability. Tensions are also apparent between incremental processes of adaptation, particularly those that focus on enhancing the resilience of existing urban systems, and those that recognize the need for a fundamental transition or transformation of urban conditions in order to address the ways in which vulnerability is currently produced in cities. Seeking to build resilience may offer a practical means of responding to the realities of urban vulnerability and the significant deficits in adaptive capacity and infrastructure provision facing many urban communities, or it may provide a means by which powerful economic and political elites continue to maintain the status quo and seek to protect only the most valuable parts of the city. Although transformative forms of urbanization may offer more radical potential for addressing such concerns, the extent to which they are likely to be adopted and implemented where they challenge vested interests and incumbent sociotechnical systems is uncertain. Understanding the dynamics of urban response to climate-change adaptation, therefore, means analysing how this complex landscape is negotiated and contested in specific urban places.

Discussion points

- How and why might we distinguish between adaptation and forms of good governance and progressive development? What are the implications of sustaining or abandoning such distinctions?
- How can we explain the relatively recent engagement of municipalities with the issue of climate-change adaptation? What are the potential advantages and disadvantages for creating municipal adaptation policy?
- For one or more case-study city, compare and contrast the emergence of adaptation strategies and measures. How far has adaptation been a 'top-down' process? What roles have communities and other actors had in the process? To what extent is adaptation following a path of resilience, transition or transformation, and how might this be explained?

Further reading and resources

The literature on climate-change adaptation at the urban level is still in its early stages. The recent book by Mark Pelling (2011a), *Adaptation to Climate Change: From*

Resilience to Transformation, provides a thorough introduction to the issues and challenges of adaptation and is the origin of the threefold framework used in this chapter to distinguish different forms of adaptation.

In addition to the core texts listed in Chapter 1, a recent special issue of *Current Opinion in Environmental Sustainability* (2011, volume 3) provides an excellent collection of up-to-date research papers examining issues of urban adaptation in different regions of the world. Specific discussions of the links between urban planning and climate-change adaptation can be found in:

Davoudi, S., Crawford, J. and Mehmood, A. (Eds) (2009) *Planning for Climate Change: Strategies for Mitigation and Adaptation for Spatial Planners*. Earthscan, London and Sterling, VA.

Wilson, E. and Piper, J. (2010) *Spatial Planning and Climate Change*. Routledge, Abingdon.

Examples and case studies of urban adaptation policies and measures can be found by searching the websites of individual cities. Some well-known examples are:

- Chicago: www.chicagoclimateaction.org/pages/adaptation/11.php
- Durban: www.durban.gov.za/City_Services/development_planning_management/environmental_planning_climate_protection/Pages/default.aspx
- London: www.london.gov.uk/lccp/
- Quito: www.quitoambiente.com.ec/index.php/cambio-climatico
- Rotterdam: www.rotterdamclimateinitiative.nl/en.

In addition, transnational networks and other actors working in the field of urban adaptation are beginning to create web-based resources, including:

- ACCRN: www.acccrn.org/
- ICLEI: http://resilient-cities.iclei.org/bonn2011/resilience-resource-point/
- UN-Habitat: www.unhabitat.org/categories.asp?catid=550.

7 Climate-change experiments and alternatives in the city

Introduction

Over the past two decades, there has been a growing interest in developing urban responses to the twin challenges of climate-change adaptation and mitigation. Such responses can be considered as forms of governing – means of intervening in order to direct or guide the actions of others (Chapter 4). For the most part, research and policy attention has been directed at the actions of municipal authorities and the roles that they have taken and could develop in order to reduce urban vulnerability and GHG emissions. Chapter 5 examined in detail the design and implementation of mitigation policy. Chapter 6 focused on adaptation and considered the ways in which municipal and other authorities had sought to pursue adaptation or to foster community-based forms of adaptation. In each case, although ideal models of policy development are emerging, in which scientific knowledge is used as the basis for deriving targets and plans that are then implemented and assessed, in practice the push and pull of various factors that both enabled and constrained urban responses are creating a more complex and fragmented picture of how cities were mitigating and adapting to climate change.

As a result, although there is evidence that comprehensive policy and planning efforts are emerging in some cities with regard to mitigation, 'numerous cities, which have adopted GHG reduction targets, have failed to pursue such a systematic and structured approach and, instead, prefer to implement no-regret measures on a case by case basis' (Alber and Kern 2008: 4; see also Jollands 2008). Similarly, in terms of adaptation, 'the absence of models to follow has led local governments pursuing adaptation to test new ideas and approaches at every step in the planning process' (Anguelovski and Carmin 2011: 171). At the same time, as climate change has become more broadly regarded as an issue of economic, political and social concern, urban responses to the issue have moved beyond the boundaries of municipalities to encompass a wide range of actors and interests (Chapter 4). The result is a rather paradoxical situation. On the one hand, as Chapters 5 and 6 demonstrate, despite sustained commitment to pursuing urban responses to climate change and their increasing strategic importance, the development and deployment of urban plans and policies remain limited. On the other hand, urban landscapes are littered with projects and schemes of all kinds that lay claim to be a form of response to climate change – from the refurbishment of derelict land to tree-planting schemes in urban parks and from dress codes for office wear in hot weather to calls to reuse water bottles, climate change is attaching itself to the strategic and mundane processes through which cities are organized and lived.

Against the backdrop of the urgency for action on climate change, such piecemeal responses may seem deeply unsatisfactory and could be taken as evidence of the lack of institutional and political capacity to co-ordinate and deliver an integrated, planned approach for urban climate governance (Corfee-Morlot *et al.* 2011). An alternative view suggests that this patchwork of interventions is critically important for understanding urban responses to climate change. To start with, the seeming ubiquity of such initiatives and schemes requires some form of explanation. Should their presence be taken as an indication that existing policy approaches have failed? Or are they a mark of the success of discourses about the need for, and significance of, addressing climate change in the city? In addition, there is a need to understand the nature and dynamics of such interventions, for they could offer the means by which different forms of response to climate change are tested and learning developed, offering the potential for broader change. Furthermore, such forms of intervention may provide spaces within which approaches to addressing climate change in the city that offer an alternative to those rooted in concerns for securing resources and pursuing 'carbon control' can emerge. In these ways, such incremental and unco-ordinated urban responses can be considered an important part of the ways in which governing climate change in the city is taking shape.

This chapter examines this urban mosaic of climate-change responses. The chapter is divided into three sections. The first considers how and why such forms of intervention might be fundamental to the ways in which climate-change governance is being conducted in the city. Rather than regarding such projects and schemes as stand-alone, one-off examples of best practice, they can be conceptualized as 'climate-change experiments' that are a critical means by which state and non-state actors seek to intervene to address climate change (Bulkeley and Castán Broto 2012a). Initial evidence suggests that such experiments are now taking place across very different types of global city, and, although they are primarily led by municipalities, they are also providing a means by which private and civil-society actors are mobilizing urban responses to climate change. The second section of the chapter considers in more detail examples of climate-change experiments. Distinctions are made between experiments as forms of policy innovation, as eco-city developments, as technical interventions and as efforts to reconfigure everyday practices, and the purposes, limitations and implications of climate-change experiments are considered. The third section of the chapter examines the emergence of alternative experiments, those that depart from the mainstream discourses of urban climate-change responses concerned with economic development, resource security and carbon control and engage with other agendas, including those of social and environmental justice and fundamental transitions in the economic and social basis of urban life. Although far from common, the presence of these forms of experimentation suggest that how cities should respond to climate change is contested, and that this remains a terrain of political struggle. The conclusions summarize the main findings from the chapter and provide suggested discussion points as well as further reading and resources.

Sowing the seeds of change?

Although advocates of municipal responses to climate change, including transnational networks, international donor organizations, national governments and municipal authorities, have advocated a sequentially organized, evidence-based approach based on the seemingly incontrovertible logic that 'if you can't measure it, you can't manage it' (C40 2011b), the realities of developing urban responses have been somewhat different. Reflecting a broader set of processes that have characterized the shifting nature of urban governance, on the one hand, and the evolving climate-change agenda, on the other, a more incremental, responsive and opportunistic approach has characterized many municipal processes of responding to both mitigation and adaptation. This reflects the institutional structures of local governance, the shifts in the power and capacities of local government that have taken place over the past two decades, and the various institutional, political and sociotechnical challenges discussed in Chapters 5 and 6. This section considers these processes in more detail, examines how the resulting initiatives and interventions can be conceptualized as climate change *experiments*, and explores initial evidence about how such climate-change experiments are emerging in global cities (Bulkeley and Castán Broto 2012a).

From plan-based to project-based urban climate governance

Among the most important factors shaping urban responses to climate change have been shifts in the very nature of urban governance. In the UK, for example, the 1980s and 1990s witnessed reforms through which 'local authorities became less extensively involved in the direct provision of education, housing, public transport, social and other services. Instead, they increasingly "enabled" other agencies, the voluntary sector and the private sector, to provide these services' (Leach and Percy-Smith 2001: 29). In Germany, shifts in the nature of local governance have also taken place: 'an important (and time-honoured) segment [of local government] seems to be breaking away, as, under the pressure of European market-liberalisation' (Wollmann 2004: 654), municipalities are withdrawing from the provision of energy and transport services (Bulkeley and Kern 2006). In the US and Australia, where municipal governments have traditionally had a weaker role, the emergence of principles of neo-liberal governance has served further to limit the general competencies for municipalities in the area of climate governance. Although the principles and effects of neo-liberalism are far from universal, they have served to shape the development of urban climate-change responses in Europe, North America and Australia. In other local-governance contexts, limited resources and capacity to address climate change, particularly in the face of other more immediate concerns, have also served to contribute to a situation in which developing integrated and strategic approaches to the issue is a significant challenge.

In this context of ongoing reform and limited capacities, and faced with an issue that cuts across traditional municipal sectors and lies outside core competencies, municipalities turned to focus on modes of self-governance and enabling (Chapter 4). The self-governing mode has been shaped by the kinds of systematic and evidence-based policy approach advocated above. However, with several notable exceptions, it has been where municipalities have sought to respond to climate-change mitigation at the level of the community or have sought to develop adaptation responses that a more ad hoc approach has emerged. In part, this reflects the very nature of enabling as a mode of governing and its dependence on what John Allen (2004: 27–8) terms 'inducement' – such as financial incentives – and 'seduction' – attempts to win hearts and minds – as well as what could be termed 'generative power, the power to learn new practices and create new capacities' (Coafee and Healy 2003: 1982). This, in turn, means that, where an enabling mode of governing is deployed, it is more likely to operate through discrete projects that provide windows of opportunity and incentives for participation. The ad hoc approach being developed in response to climate change is also a reflection of the political work undertaken to reframe or localize climate change in relation to other issues – whether this is in terms of financial savings or disaster relief – and to utilize specific windows of opportunity as a means of advancing climate-change action. Although this has proven to be a valuable strategy for ensuring the place of climate change on local agendas, it serves to reinforce an opportunistic, case-by-case approach to the development of initiatives and measures (Bulkeley and Kern 2006; Sanchez-Rodriguez et al. 2008). Such interventions have also been affected by the availability of funding, which is often shaped around the delivery of specific projects and programmes and is often short-term in nature, further contributing to a patchwork of urban responses. The reinvigoration of transnational municipal co-operation in the early 2000s, through, for example, the work of C40, ICLEI and various mayors' agreements (Chapter 4), has both been a response to this context and has served to perpetuate it by leveraging project-based funding and focusing on a range of enabling actions.

These factors go some way towards explaining why municipal authorities have followed an incremental, responsive and opportunistic pathway to the development of mitigation and adaptation policies and measures. The emergence of urban responses to climate change has not, of course, been confined to municipal actors. A range of other organizations, from funding bodies and national governments to private companies and community-based organizations, have also sought to address climate change through projects and initiatives that focus on one or more cities. Given that many such actors, by themselves, lack the potential for effecting systemic or strategic interventions in particular cities, either because they operate across multiple urban arenas or because they lack the power and authority to intervene in such a manner, it is perhaps unsurprising that the sorts of intervention that they have been concerned with focus on particular buildings, demonstration projects or community-based initiatives. At the same time, as climate change comes to gain in popularity and take a place in public culture, various products, sites, measures and initiatives, which would previously have been developed in the absence of climate change, have come to be associated with the issue. The malleability and ubiquity of climate change as a discourse mean that it can become attached to multiple projects, from flood-protection measures to tree-planting schemes, adding to the fragmented landscape of urban responses (Bulkeley and Castán Broto 2012a).

Understanding climate-change experiments

In much of the literature on urban responses to climate change, and indeed urban governance more broadly, such initiatives and interventions are regarded as one-off or best-practice projects, somehow separate from the real business of planning and governing cities. However, as the discussion above makes clear, they can also be seen as the *outcome* of governance as usual and, in this way, an essential part of the way in which governing is conducted. Several different bodies of literature have drawn attention to the significant role that forms of innovation or experimentation have in the making of policy, and in the process of governance more broadly. In the early twentieth century, the American lawyer Louis Brandeis famously argued that the US states function as 'laboratories of democracy', by 'testing new ideas and policy proposals, gradually building a record of policy innovation that can be tapped by national officials when the time is ripe' (Aulisi *et al.* 2007: 5). Rather than being the antithesis of policy and planning, therefore, experimentation and innovation can be seen as integral to the ways in which policy is formulated, and subnational government can be seen as as a particular arena within which this takes place.

More recently, Matthew Hoffmann (2011) has persuasively argued that experimentation is a critical part of the way in which climate governance is evolving. He suggests that, 'climate governance experiments' are emerging across different political jurisdictions (transnationally, regionally), as increasing frustration with the international processes of negotiation and climate agreement, coupled with fragmentation of political authority across a diverse range of public and private actors, creates the political space within which alternative governance arrangements and initiatives can emerge. As forms of governance, these are efforts to get things done, to make rules, in the absence of formal authority; they are experimental insofar as they are 'innovative' and imply 'trial and error' with novel governance mechanisms beyond the process of multilateral treaty making (Hoffmann 2011: 17). For Hoffmann, climate-governance experiments are defined by three criteria: they explicitly seek to make rules or norms that 'shape how communities respond to climate change'; they are independent of the international process of climate governance or national regulation; and they cross jurisdictional boundaries (Hoffman 2011: 17–18). This third criterion, he suggests, is necessary in order to restrict experiments to those that are 'are rule-making endeavours in non-traditional political spaces' (Hoffmann 2009: 4). Given that many urban responses to climate change cut across existing institutional channels for making and implementing 'rules', both within and beyond municipalities, some urban initiatives that do not cross jurisdictional boundaries could also be included within this definition.

Drawing on Hoffmann's work, therefore, urban interventions, projects and initiatives that seek to develop rules, in the broadest sense, to govern the actions or conduct of others and that take shape beyond well-established channels of decision-making could be regarded as forms of 'climate-governance experiment'. Such experiments, as Hoffman argues, are also characterized by a more or less explicit intent to innovate, to foster some form of learning or to gain new experience about the potential for governing climate change. However, rather than being driven solely by the changing international climate-governance context and different forms of motivation, such as profit, a sense of urgency, ideology or efforts to secure resources, as the discussion above makes clear, urban climate experimentation is also shaped by the dynamics of political economy and authority taking place within different cities.

Hoffmann's analysis provides a useful starting point for considering 'experiments', not as stand-alone initiatives, but as part of the broader governance landscape. Other conceptual approaches have also pointed to the wider role that experiments can play. Literatures that seek to explain the dynamics of sociotechnical regimes – a term used to describe the social and technical elements that constitute and co-produce infrastructure systems and that lead to the dominance or lock-in of particular systems – suggest that 'niches' and experiments can provide a catalyst for change (Geels and Kemp 2007; Smith *et al.* 2010). Technological niches, for example, can be 'made operational through (a series of) protected test beds such as pilot and demonstration plants where technologies are applied in a societal setting for the first time' (Raven 2007: 2391). Others have pointed to the importance of 'grass-roots innovations' as niches in incumbent sociotechnical regimes, 'bottom up experiments with environmental technology by citizen groups and/or NGOs, operating outside the institutional structures of firms and governments' (Hegger *et al.* 2007). In each case, niches are regarded as offering a protected space within which new ideas can be generated and new technologies, forms of social organization or practice can be trialled. In contrast to Hoffmann's account, here, experiments are important because they intervene in social *and* technical systems.

Work that examines specific urban laboratory projects makes a similar argument, suggesting that it is because of the conjunction of the social and technical that undertaking live experimentation in cities is becoming part of urban-governance strategies. In his analysis of the emergence of so-called living laboratories – research projects designed to test particular forms of sustainability intervention with social and natural urban systems in real places and real time – James Evans (2011) documents how such experiments are associated with a 'particular style of adaptive governance that seeks to feed environmental monitoring back into a management process' (Evans 2011: 255). As cities come to be regarded as self-regulating, socio-ecological entities, and processes of governance shift from those focused on managing to steering, such forms of adaptive governance are advocated as a new means by which to shape urban futures in the context of uncertainty. Indeed, such forms of flexibility, adaptiveness and experimentation are regarded as essential features for governing sustainability. For example, in the context of urban water governance in Australia, Farrelly and Brown argue that, 'sustainable regimes would emphasize an adaptive framework, prioritizing flexible, inclusive, and collaborative practices, operating within organizational cultures that embrace experimentation and learning to foster sectoral adaptation' (Farrelly and Brown 2011: 721). Experimentation as living laboratory is, therefore, not only about fostering scientific learning, but also about a new way of governing the city.

Literature on climate governance, sociotechnical regimes and the emergence of living laboratories in cities all points to the ways in which experimentation is tied into broader processes of governing the city. This suggests that the term 'climate-change experiment' is a useful one to frame the sorts of intervention and initiative that are emerging in cities that seek explicitly to develop new rules, foster learning/experience, transgress existing institutional boundaries and reconfigure sociotechnical relations. Not all initiatives and interventions will have this experimental quality. Some may emerge from tried and tested means of decision-making. Others may not be concerned with developing innovation or learning. Those that do, however, may be particularly significant, because of their potential catalytic effect within sociotechnical regimes, and because they serve as new means by which governing is conducted in the city.

Mapping urban climate-change experiments

During 2009–10, a research team at Durham University, UK, conducted a survey of 100 large cities worldwide as part of the Urban Transitions and Climate Change (UTACC) project to examine the nature and extent of climate-change experimentation taking place.[1] As the first survey of its kind, it provides a valuable source of evidence about where and when experiments have emerged in the urban arena, as well as the sectors in which they are taking place and the actors who have been involved in this process of experimentation.

The survey found that 79 per cent (495 experiments) started after 2005, that is, after the Kyoto Protocol was ratified. Only 5 per cent of initiatives started before its initial adoption in 1997 (Bulkeley and Castán Broto 2012a; Castán Broto and Bulkeley 2012). This finding is in keeping with Hoffmann's analysis that climate-governance experimentation is a relatively recent phenomenon, emerging as both the Kyoto Protocol created opportunities for new actors to intervene in climate governance (e.g. through the financial mechanisms, such as carbon trading and the CDM that it contains) and its subsequent negotiation led to further concerns about the viability of an internationally negotiated resolution to the climate-governance problem. However, Hoffmann's analysis has focused on experiments at the transnational and regional scale. As discussed above, there are some specific features of the changing landscapes of urban political economies, and the ways in which climate governance at the level of the city has evolved, that might also explain the recent emergence of urban climate-change experiments. These include the growing enabling role of municipalities, the rapid urbanization characteristic of cities in the developing economies, and the range of actors who are increasingly regarding the city as a site of climate action. The UTACC survey found that experimentation was not confined to cities in North America, Europe and Oceania, where climate governance has historically been concentrated, and instead that the distribution of experiments in each global region was in accordance with the number of cities included in the survey from that region (Bulkeley and Castán Broto 2012a; Castán Broto and Bulkeley 2012).

In terms of the types of sector and forms of response that are taking place in relation to mitigation, the survey found that a focus on urban infrastructure (excluding mobility) was the commonest, followed by the built environment and transport (treated in the survey as a separate category), and lastly urban-development sectors, planning and urban greening or carbon-sequestration experiments (Figure 7.1). Mirroring previous research findings on urban responses to climate change, and the discussion in Chapter 6, the survey found considerably fewer experiments that specifically addressed climate-change adaptation (Figure 7.1; see also Bulkeley and Castán Broto 2012a; Castán Broto and Bulkeley 2012). Overall, 45 per cent of experiments focused on the energy sector. In the infrastructure sector, 78 per cent of experiments were energy-related. The ongoing association between energy saving and financial gain, together with new market opportunities in cities, including forms of low-carbon investment and finance and with emerging forms of carbon control (While *et al.* 2010), may help to explain this focus (Bulkeley and Castán Broto 2012a). The types of innovation and experimentation that are taking place are both social and technical. The UTACC survey found that 76 per cent of experiments focused on technical innovation, and social innovation was

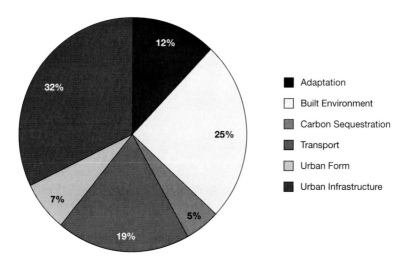

Figure 7.1 Distribution of experiments in different sectors in the UTACC database

present in 50 per cent of all experiments (Castán Broto and Bulkeley 2012). This varied between sectors. In the infrastructure sector, 88 per cent of experiments had a technical-innovation component, but only 39 per cent had a focus on social innovation. In the carbon-sequestration sector, in contrast, only 40 per cent had a technical-innovation component, and 60 per cent had a social-innovation focus (Castán Broto and Bulkeley 2012).

In addition to analysing when, where and how experiments are taking place, the UTACC survey also sought to understand who is undertaking experiments and how they seek to govern climate change in this manner. The survey found that municipal authorities, a category that in this case included some municipal agencies that undertake utility provision or provide transportation, were the dominant actor, leading 66 per cent of experiments (Castán Broto and Bulkeley 2012). Other actors were also important initiators of experiments, with private actors leading 15 per cent of initiatives, and civil-society actors leading 9 per cent of them (Castán Broto and Bulkeley 2012). However, many experiments were not organized and implemented only by one actor, but instead were conducted in partnership. Almost half of the experiments (47 per cent) involved some form of partnership, either vertically, between different levels of government, or horizontally, between different urban actors (public, private, civil society) (Castán Broto and Bulkeley 2012). This is important, because it shows that, although municipal actors may continue to dominate, private actors and civil-society actors are playing an

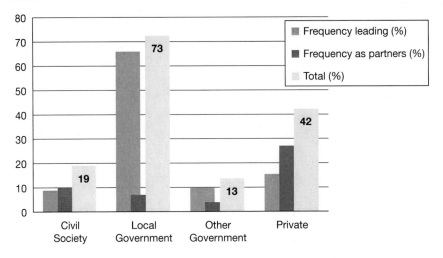

Figure 7.2 The frequency with which actors participate in climate-change experiments in the UTACC database

increasingly recognized role in urban climate governance. Although these actors may lead only a quarter of all initiatives between them, private actors participate in 42 per cent of experiments, and civil-society actors participate in 19 per cent of them (Figure 7.2).

In their relatively recent origin, urban climate-change experiments reflect a key feature of the broader phenomenon of governance experiments that Hoffmann has identified. However, the concern of these urban climate-change experiments with forms of technical and social innovation suggests that they can also be understood as new niches or laboratories opening up, as cities seek to develop new ways of understanding or experiencing the city in relation to climate change. Municipal actors have been engaged with much of this activity; however, other actors, and in particular the private sector, are increasingly formally recognized as an important part of this form of urban climate governance, lending weight to the argument that the nature of urban climate governance is changing (Chapter 4).

Experiments in practice

As well as the emergence of urban policy and planning for climate-change mitigation and adaptation, there are, therefore, a range of other responses taking place in cities that take the form of projects and interventions. Some of these interventions can be considered as forms of experiment – means of seeking to understand and experience the city differently in relation to climate change. Such urban climate-change experiments have emerged predominantly since 2005, are not confined to particular cities or regions, and involve both technical and social innovation. This section examines how experiments operate in practice. Focusing on four different types – policy innovations, eco-city developments, innovative technologies and the transformation of everyday practices – it considers the purposes, potential and limitations of experiments.

Policy innovation

In some senses, urban climate-change experiments that focus on policy innovation may come closest to Hoffmann's definition of climate-governance experiments. Here, experiments have been designed to establish new ways of organizing urban climate policy, as well as to develop new policy tools and instruments. Hoffmann (2011) identifies city networks as one form of transnational governance experiment, finding that, of the fifty-eight governance experiments that he studied, '14 are at least partially engaged in technology deployment in cities'. According to Hoffmann, this forms a distinct cluster of governance experiments that includes those transnational urban-governance networks which specifically focus on climate change and which encourage members to develop and deploy new forms of technology (e.g. ICLEI, C40), 'a range of technology focused experiments (Climate Neutral Network and Connected Urban Development Program) that find municipalities and their networks a fruitful level of political organization in which to deploy their work', as well as governance experiments that 'seek to bring cities and corporations together to work in this area', including, for example, the Climate Group and Clinton Climate Initiative (Hoffmann 2011: 108). In this manner, policy innovations taking place transnationally, and operating between public and private actors, are leading to the development of new responses to climate change in cities. Hoffmann's analysis usefully shows the increasing connections and overlaps between these initiatives, and the ways in which they interact, both in order to co-operate with one another and to compete. This may lead to new synergies and divisions of labour emerging, where some networks become specialized, but it may also lead to a confusing array of different policy approaches and technological strategies on offer, with limited capacity for cities to select and deploy appropriate approaches (Hoffmann 2011: 120–2).

In addition to the forms of policy innovation that have emerged between cities, there is also evidence of urban climate-change experimentation that focuses on policy innovation within the urban arena. Such policy innovations have included the development of new governance arrangements – such as climate-change commissions or think tanks, which provide interdisciplinary and cross-sectorial assessments of urban climate-change challenges and opportunities – the development of new tools and methodologies for evaluating projects, including carbon-footprint analysis, and the use of new tools or policy instruments, most notably those that are related to carbon and emissions trading (Box 7.1).

One thing that marks out such forms of experimentation is their transgression of traditional divides within municipalities and between the public and private spheres. One example of how this takes place in practice can be found in Pennsylvania. Although inventories of GHG emissions are nothing new, there remains an absence of standard approaches and measures for inventorying emissions beyond the municipality itself (Hillmar-Pegram *et al.* 2011: 78). In Pennsylvania's Centre Region, a group of researchers from Penn State University embarked on an inventory process in order to inform local policymakers and provide the basis for the creation of mitigation strategy. This process involved the creation of an emissions inventory, which involved making decisions about the scale and scope of analysis, and a further set of deliberations with local communities and stakeholders. In stage one of these deliberations, community focus groups engaged residents, who were informed by the emissions inventory, in the generation of possible mitigation options for the region. Stage two involved communities and stakeholders prioritizing these options, and, in the third stage, 'the local government uses the publically vetted mitigation options as a basis for considering a formal action plan' (Hillmar-Pegram *et al.* 2011: 81). In this manner, the research team intended to engage both scientific and community knowledge about the potential options for addressing climate change in the region, and to use this as a basis for informing policy decisions. However, once the mitigation options that had been deliberated upon by community and stakeholder representatives were discussed with policymakers, disagreement emerged among the policy community as to whether climate change was a real phenomenon that justified local action, and where any such responsibilities might lie (Hillmar-Pegram *et al.* 2011: 83). This created an impasse within the policy system, which was eventually overcome through extended deliberation and the persistence of a few key individuals, and a switch in focus from climate change to energy efficiency. The final 'Model Resolution 2011–1 . . . contains 33 "Energy Efficiency Action Items" that were drawn directly from the climate change

BOX 7.1

Urban emissions trading experiments

Emissions trading schemes (cap-and-trade) are not new in environmental economics. Since the 1960s, economists have been modelling how market mechanisms could address externalities such as environmental pollution. What is new, however, is the prominence that this idea has been given within climate-change debates and how it has been adopted as a flagship policy in some global cities.

One of the most prominent emissions trading markets was the Chicago Climate Exchange (CCX), which operated between 2003 and 2010 as North America's sole voluntary, legally binding emissions trading arena. CCX also provided opportunities to invest in carbon-offset projects in different parts of the world. The CCX was inspired by Richard Sandor, an American economist who first implemented a cap-and-trade programme for sulphur dioxide (SO_2) emissions, with the intervention of the government putting a cap on emissions and forcing companies to trade emission rights. Following this experience, and with funds from the Joyce Foundation, the CCX operated until 2010 to exchange trades in emissions of six GHGs: carbon dioxide, methane, nitrous oxide, sulphur hexafluoride, perfluorocarbons and hydrofluorocarbons. The CCX was run by a public company, Climate Exchange PLC. It consisted of a trading platform (to execute the trades among members) and a clearing and settlement platform (which executed the transactions). It also had an official database of Carbon Financial Instruments of members, or Registry.

The CCX, however, ended trading in November of 2010 after Intercontinental Exchange Inc., a financial company, acquired Climate Exchange PLC in July 2010. The two major problems in the CCX were the continuous fall of carbon prices (from more than US$7 in 2008 to 10 cents per tonne in 2010) and the lack of political momentum in Washington to pass a cap-and-trade bill (both because of the Republican majority in the US House of Representatives and the lack of support for the bill from Democrat senators in industrial states). Nevertheless, as an experiment in carbon governance, the CCX attracted 450 heterogeneous members, including power companies, cities, universities and big corporations such as Ford, DuPont, Motorola, International Paper and

Honeywell. The CCX calculated that it had reduced carbon dioxide by 700 million tonnes since 2003 (88 per cent in reductions in industrial emissions and 12 per cent from offset projects).

Before the collapse of the CCX, the criticisms raised against it related to either the potential of cap-and-trade mechanisms to bring about the radical change that is envisaged in a low-carbon society (in a mechanism that is essentially designed to make small, incremental improvements) or the ethical problems associated with putting a price on the collective responsibility to reduce carbon emissions, which results in the transformation of an ethical, social and environmental problem into a merely economic problem. The collapse of the CCX, however, draws attention to the greatest problem, which is the difficulty of making trading agents believe they can actually obtain a benefit from trading emissions. In some respects, the case of the CCX highlights that climate change cannot be conceptualized as a mere externality to existing carbon markets, but is a central issue that poses ethical questions regarding the functioning of our societies, the translation of unequal social and economic relations into a global problem, and the collective responsibilities for new anthropogenic risks.

The CCX was an initiative led by a group of public and private actors, who used Chicago – the city – as a platform to reach potential members in North America. Other emission trading exchanges, however, have been supported and promoted by local governments, with the aims both of raising the profile of the city and of tapping into the potential wealth that carbon trading is thought to generate. This is the case for Tokyo, where there is a pioneering cap-and-trade initiative led by the Tokyo Municipal Government (TMG). In 2005, a Metropolitan Security Ordinance (the CO_2 Emission Reduction Programme) required business establishments considered as 'large emitters' to submit a five-year GHG-reduction plan, which would be evaluated with a system of ratings (A, A+, AA). The Tokyo government identified the cap-and-trade system as a direct means to step up emissions reductions, not only by promoting the reduction of emissions of large businesses but also by enabling them to purchase emissions reductions from smaller business. The proposal by the Tokyo government is qualitatively different from the CCX, because, in Tokyo, the local government enforces the cap-and-trade mechanism (and it does not depend on political debates at the national level to do so). This means, however, that its jurisdiction has a more limited scope, and, thus, the value of transactions is less than in a national-level market. The government heralded this as 'the world's first urban cap-and-trade programme' when it started operating in April 2010, just as Intercontinental Exchange Inc. announced the fate of the CCX. In 2009, the TMG had made broader proposals for the implementation of a nation-wide cap-and-trade programme, which could reach larger suppliers outside the jurisdiction of the metropolitan government and take advantage of existing carbon-sequestration programmes nationwide.

There are 1,340 businesses participating in the scheme, and, according to some observers, it has been accepted among stakeholders (Padeco 2010). The TMG establishes targets for emissions reductions, and fines are imposed on businesses that do not meet those targets (which include monetary fines of up to ¥500,000 (US$5,500), additional obligations and the publication of the names of offenders, as a form of 'shaming the offender'). To meet the targets, businesses can either reduce their own emissions independently or trade credits, including: excess reductions from other companies; credits from small and medium enterprises (now subject to fines); credits from facilities outside the city (only up to a third of year emissions); renewable energy credits or credits from the City Solar Energy Bank (TMG 2008). Rather than relying on emissions reductions, the Tokyo exchange has a clear regulatory approach to the cap-and-trade system.

The same reservations expressed in the CCX case, about the suitability of a cap-and-trade emissions programme, apply in the case of Tokyo. However, it appears that the clarity achieved at the local level – with a direct correlation in terms of where the emissions reductions occur and what activities are financed through cap-and-trade – makes the mechanism more fungible, and, thus, more stable from a financial perspective. Moreover, marketed as the first urban cape-and-trade mechanism in the world, the city shows leadership as a global city and engages with renewed discourses of environmental conservation in the city and a sense of global responsibility. Other cities, however, have gone for the potential profitability of voluntary markets at the national level (such as, for example, the Tianjin Climate Exchange, which has followed the CCX model), acting more as a host of the marketplace than as a regulatory body.

Vanesa Castán Broto, Development Planning Unit,
University College London, UK

mitigation options generated and prioritized by the focus groups', and, in 2011, these actions were being incorporated into municipal work plans (Hillmar-Pegram *et al.* 2011: 83).

Eco-city experiments

A second type of urban climate change experiment focuses on various scales and forms of eco-city development. Experimentation with different forms of 'eco-city' can be traced back to the early-twentieth-century movement for 'garden cities' in the UK, through the development of Arcosanti in Arizona in the 1970s, to more recent examples such as Masdar City in the UAE. A recent survey found that

> innovative eco-city initiatives are as likely to be found in China, Kenya, Japan, South Korea, and South Africa, as in Canada, Germany, Great Britain, Sweden, and the United States. Some of the most original eco-city projects are currently in planning or under construction in the Middle East and East Asia.
>
> (Joss 2010: 242)

Whereas eco-city developments can have different sustainability concerns, from the reduction and reuse of wastes to water conservation, the conservation of resources to new forms of economic activity, a range of street, neighbourhood and even city-scale eco-developments have emerged over the past decade that explicitly seek to address climate change.

One example of eco-development as urban expansion can be found in Bangalore, India. Here, the T-Zed (Towards Zero Carbon Development) project, targeted at the higher-income residents of the city, has been developed as an innovative, low-carbon residential development on the urban fringe. Led by a private company, BCIL (Biodiversity Conservation India), like many other developments in this area, it is a gated community or compound. The development includes sixteen single-family houses and seventy-five apartments and incorporates numerous forms of social and technical innovation, focused on reducing carbon and limiting reliance on wider systems of water and energy provision (Bulkeley and Castán Broto 2012b). These include the use of low-carbon materials, renewable-energy systems, water-saving measures and efforts to change the behaviour of residents.

T-Zed has created a space for climate-change innovation, through the replication of T-Zed-like development projects, seeding new companies (e.g. Flexitron, a light-innovation company), and the uptake of some of its principles and approaches in policy at local and national levels. At the same time, both through principles of autonomous development and through its eventual reliance on borehole water, T-Zed also reconfigures wider urban, sociotechnical networks. It has provided a site within which new discourses take place about what low-carbon living might entail for middle-class residents in Indian cities, messages that have been taken to neighbouring developments and have become part of wider urban discourses. This has not been a straightforward process, and many elements of T-Zed have been contested. At the same time, although such low-carbon developments hold the promise of creating new ways in which urban expansion might be managed, they also serve to replicate forms of low-density, urban sprawl that might increase reliance on private motorized vehicles and that might ultimately sustain processes of urban development in which those who currently suffer urban poverty and exclusion continue to do so.

Novel technologies

As outlined above, the majority of climate-change experiments have a dimension of technical innovation. Some experiments are more specifically designed to develop and test new technologies, and these can range from small-scale applications of technologies that may be reasonably well known in one urban context but have yet to be trialled in another, to large-scale applications of technological systems that have yet to be tested at scale.

In Australia, Farrelly and Brown (2010) examined eleven experiments in the water sector concerned with securing alternative forms of water supply and protecting water-system health in the face of changing climatic conditions and a growing demand for water services. Although these experiments exhibited a mixture of what they term 'structural (i.e. technology and infrastructure) or non-structural innovations (i.e. policy program)', most focused on 'a new scale of technology (decentralized) for the provision of alternative water sources (i.e. recycled water) for non-potable use' (Farrelly and Brown 2010: 724; Figure 7.3). Many of these technologies are not experimental in the sense of being novel or untried, but, rather, where their experimental quality comes from is in their application in particular contexts, the purposes for which they are being used, and in terms of the actors who are seeking to deploy these approaches. In their work, Farrelly and Brown (2010: 729) found that, 'local-scale experiments are valuable learning platforms for urban water practitioners, but their legitimacy and ability to influence the regime is significantly impeded by the systemic inertia of traditional urban water management practices'. In other words, although experiments may offer the kinds of learning and way in which new forms of living in the city under conditions of climate change can be experienced, the extent to which they can shape existing and embedded ways of, in this case, managing water in the city is moot.

Such technological innovations do not only take place as small-scale, local projects, but can also be tied to wider agendas and to the activities of a range of established political and economic actors. National drivers for energy security, coupled with the international climate-change regime, have led to

Table 7.1 Examples of climate-change experiments in the Australian water sector

Project	City	Innovation/technologies
60L Green Building	Central Business District, Victoria	Onsite black- and grey-water treatment Rainwater harvesting Water-saving devices
Aurora Estate	Epping North, Victoria	Decentralized wastewater recycling plant Dual-pipe reticulation to houses Storm-water quality treatment train Rainwater tanks for hot water
Inkerman Oasis	St Kilda, Victoria	Onsite grey-water treatment and reuse for toilet flushing and garden irrigation Storm-water quality treatment
Wungong Urboan Water Project	Armdale, WA	Collaborative planning approach to integrating sustainable water design features into a master-planned development
Rocks Riverside Park	Seventeen Miles Rocks, SEQ	Passive, onsite sewage treatment (through reed-bed filters) for public open-space irrigation

Source: Adapted from Farrelley and Brown 2011: 725

increasing interest in developing new forms of energy-supply system that are low carbon. In the US, the Department of Energy has established the *Solar American Cities* programme (Box 7.2). In Boston, for example, 'in 2008, with support from the Solar America Cities grant, Boston city government formed Solar Boston, a 2-year initiative to increase solar energy installations in Boston by a factor of 50' (Solar American Cities 2011: 2). The Solar Boston initiative has led to the development of a strategy for implementing solar technology, 'including mapping feasible locations, preparing a permitting guide, and planning the citywide bulk purchase, financing, and installation of solar technology' (Solar American Cities 2011: 3). At the international level, the CDM has proven to be one means by which cities have sought to develop alternative energy technologies, in particular energy-from-waste projects,

> including 'Aterro Bandeirantes' and 'Aterro San Joao' in São Paulo, Brazil; the Methane Plant Zambiza in Quito, Ecuador; the Biogas Capture Bordo Poniente in Mexico City; and in South Africa, the Bellville South landfill site in Cape Town and the gas-to-energy project in Johannesburg.
> (UN-Habitat 2011: 99)

As these examples illustrate, urban climate-change experiments can be conducted by actors operating at different scales, who form 'vertical' partnerships through which to implement and test new forms of response to climate change. Indeed, such forms of vertical integration, supported by new financial flows, are important aspects of such experiments.

Transformative practices

Changing the behaviour of individuals and organizations has also been an important form of experimentation. As discussed in previous chapters, where municipalities lack the direct power to regulate or provide services, the enabling mode of governance has become a critical means through which the governing of climate change takes place. Although municipalities have some established channels through which this can be pursued – most notably through education campaigns and other forms of media communication – seeking to shift patterns of behaviour has been one area in which both innovative approaches and novel techniques are emerging. Furthermore, such initiatives are not only confined to municipal actors, but increasingly involve a range of corporate and civil-society organizations that seek to address climate change through enabling changes in behaviour, everyday practices and new forms of social organization.

> BOX 7.2
>
> **About the Solar American Cities programme**
>
> Through the Solar America Communities effort, the U.S. Department of Energy (DOE) is working to rapidly increase the use and integration of solar energy in communities across the country.
>
> DOE recognizes the important role of local governments in accelerating widespread solar energy adoption. Cities and counties are uniquely positioned to reduce global climate change, strengthen America's energy independence, and support the transition to a clean energy economy by converting to solar energy sources on the community level.
>
> Through the DOE Solar America Communities activity, local governments are working to accelerate the adoption of solar energy on the local level. The activity is based on DOE's three-pronged approach to identifying and overcoming barriers to urban solar implementation, then sharing lessons learned and best practices to facilitate replication across the nation.
>
> The original Solar America Cities Partnerships were cooperative agreements between DOE and 25 large U.S. cities established in 2007 and 2008 to develop comprehensive, city-wide approaches to increasing solar energy use.... Solar America Cities Special Projects, funded in 2009 through the American Recovery and Reinvestment Act, tackle key barriers to urban solar energy use that were identified through the 25 city partnerships.... Solar America Communities Outreach Partnership is an effort to share the best practices developed through the partnerships and special projects with hundreds of other local governments, accelerating solar energy adoption across the United States. Learn more about the outreach partnership.
>
> (Solar American Cities 2012)

Across the UK, many municipalities have organized new forms of campaign to engage and enable members of the public to make changes to their behaviour that reduce GHG emissions. What is common across initiatives such as 'Oxford is My World' (Box 7.3), 'Manchester is My Planet' (see Chapter 5) or the adoption by Middlesbrough of 'One planet living' is the development of new coalitions of actors, from within the municipality and beyond, that are seeking collectively to inform and demonstrate how individuals can make changes to reduce GHG emissions and reduce vulnerability. In London, multiple community-based groups have been established that are seeking to develop and support behavioural change. In Balham, south London, for example, the Hyde Farm Climate Action Network have taken part in the EU-funded ECHO Action project to enable up to twenty-five households to reduce their carbon dioxide emissions by 10–20 per cent, and established a Green Streets project with the energy provider British Gas, including up to forty households in a suite of measures including draught proofing, energy audits, the installation of renewable energy, and the training of local residents to receive formal qualifications in energy efficiency.[2]

BOX 7.3

Oxford is My World

Your guide to saving the planet!

'Oxford is My World' is an award-winning local project designed to help people make choices that benefit both our local and global environments by reducing our greenhouse gas emissions. We have been working with local community groups, businesses and environmental organizations to gather information on how to reduce energy and natural resource use in Oxford. The Guide includes sections on environmentally friendly Energy, Food, Lifestyle, Recycling, Travel and Water. It is divided into 'very easy', 'fairly easy' and 'not-quite-so-easy steps'.

So whatever your circumstances, if you're going shopping, doing home improvements or even getting married, consult 'Your guide to saving the planet' first so you can do the right thing for Oxford . . . and the world!

(Oxford is My World 2012)

In Hong Kong, efforts to reduce energy demand in the home have also emerged from beyond the municipality. Since 2006, Friends of the Earth Hong Kong has organized a 'Power Smart' contest to engage households, businesses and property-development companies in reducing their energy use (Figure 7.3). Once signed up to the competition, participants are provided with advice about how to reduce energy, and, at the end of the allotted period, those who have achieved the most savings are awarded either financial prizes (in the case of households) or certification of their achievements (for the businesses and property-development companies) (Friends of the Earth Hong Kong 2011). Also in Hong Kong, WWF has developed the Climateers programme, consisting of a website where individuals can find out information about how to reduce their energy use, download carbon-calculator apps and make personal pledges, as well as an ambassador scheme, which is designed to facilitate learning about the impacts of climate change in the Hong Kong area and to use social networks to generate more interest and commitment to addressing the issue (WWF Hong Kong 2012). Such schemes are not confined to households alone, and WWF Hong Kong has also developed the Low-carbon Office Operation Programme (LOOP) and Low Carbon Manufacturing Programme (LCMP) as means of working with business in the city and greater urban region to reduce energy use. In Durban, schemes to work with local businesses have been initiated by the municipality, which has established two energy-efficiency clubs with funding from the Danish International Development Agency. Research found that, through these clubs,

> participants were introduced to techniques for energy management and auditing, monitoring and targeting, carbon footprint calculations, and making power conservation plans. Members who implemented efficiency measures reported savings of up to R220 000 (US$28,000) for the 1st quarter of 2009, and the concept of 'clubs' was generally well received by the industries.
>
> (Aylett 2011)

The limits and implications of experimentation

Climate-change experiments are emerging across a range of domains in the city – policy innovation, new urban planning, novel technologies and the transformation of social practices. Their often small-scale, short-term nature has led many in the academic and policy communities to consider the critical issue of how they might be scaled up. Given the enormity of the challenge of climate change, can such initiatives really make a difference? What are the factors that foster or prevent the approaches developed and lessons being

Figure 7.3 Advertisement for PowerSmart at the Star Ferry Terminal, Hong Kong
Source: Harriet Bulkeley

learnt in these experiments becoming mainstream? Table 7.2 provides some indications of why it can be difficult for experimental projects and initiatives to enter into mainstream policy and practice. These are valid concerns, of course, but they also downplay the way in which experimentation is already part of the mainstream response to climate change – experimentation has become a critical means through which the governing of climate change is pursued and accomplished by a range of politically and economically important urban actors. Seen from this perspective, the challenges of further developing the forms of sociotechnical innovation found in experiments are not a matter of building capacity or enhancing learning, but of addressing the social, political and economic structures that both sustain experimentation and that seek to preserve mainstream responses to climate change.

Table 7.2 Challenges facing the mainstreaming of experimentation

Challenges to mainstreaming experimentation	*Explanation*
Rules and regulation	Mandatory requirements often lag behind innovation and can prevent novel technologies and approaches being applied
Policy direction and political support	In an emerging policy sphere, multiple agendas and limited political support can prevent the uptake of innovation
Economic conditions and finance	Novel technologies can appear costly, with limited knowledge of the return on investment; alternative strategies for achieving behavioural change can appear relatively low cost and enhance/attract support
Informal capacity building	Understanding of the potential of new approaches and technologies, and development of the networks and partnerships through which they emerge
Organizational culture and learning	A lack of adequate space and time for reflection, learning and engagement with the lessons from existing initatives can limit their wider uptake
Scarcity and crisis	Experimentation can thrive in the context of scarcity and crisis, where traditional ways of approaching problems are jettisoned because of the need to respond, but long-term, persistent crises can limit resources, capacity and learning, and undermine the extension of succesful experimentation

Source: Adapted from Farrelly and Brown 2011: 728

The links between experimentation and mainstream urban climate-change responses can be clearly seen in terms of the sorts of discourse and approach that are advocated. For the most part, experiments sustain the logics of co-benefits, urban ecological security and carbon control that underpin the municipal voluntarism and strategic urbanism, which underpin urban climate-change responses (Chapter 4). In other words, urban climate-change experiments reinforce notions of the additional economic and environmental benefits that can accrue from responding to climate change, and adopt logics of security – particularly in terms of planning urban places that can be self-sufficient in resource terms – and of controlling carbon in order to achieve wider political objectives and to realize the potential economic dividend of doing so. In his work on climate-governance experiments, Matthew Hoffmann (2011: 39) similarly found that they were based on the 'compromise of liberal environmentalism', in which environmental protection is predicated on the continuation of the existing liberal economic order (see Bernstein 2001). In this manner, 'experimentation is *not* a revolutionary challenge to established governance mechanisms' (Hoffmann 2011: 40), but rather can be regarded as a means through which mainstream discourses and approaches to governing climate change in the city are realized.

Alternatives: moving beyond the mainstream?

There is, however, scope for thinking about experimentation differently. Alongside experiments that serve to foster the sorts of urban ecological security and practices of carbon control that mark the recent move towards strategic urbanism, and those that offer the means through which municipal voluntarism can be enacted, alternatives can be found. In these alternatives, experimentation is used as a means for advancing different ways both of framing the climate-change problem and of considering what an appropriate response might involve. For the most part, these alternatives have emerged through forms of 'grass-roots' innovation (Seyfang and Smith 2007), but this is not a necessary condition. This section considers three different sorts of alternative form of experimentation: those that offer some form of hybrid accommodation between mainstream discourses and more radical concerns, particular with issues of environmental and energy justice; those that have emerged from alternative conceptualizations of what it might mean to be climate resilient, and particularly the 'transition-town' phenomenon; and those that are based on more radical accounts of what it might mean to live sustainably.

Hybridizing climate change for social justice

One means by which climate-change experimentation is taking place in cities is the hybridization of climate change with other concerns that are more marginal on current policy agendas. This has been particularly important in establishing new responses that explicitly address communities and households which experience some level of socio-economic hardship, and which may both be vulnerable to the impacts of climate change and may experience further economic challenges as society seeks to reduce GHG emissions through the imposition of financial measures that raise the cost of energy. Such initiatives focus on residents who experience a degree of energy poverty or vulnerability, in terms of their ability to afford the basic services that energy provides (e.g. heating, cooling, power, cooking, cleanliness). Although such programmes and initiatives have a long history and are relatively common in the UK and in some parts of North America, Europe and Australia, it has only been recently that such initiatives have sought to consider how energy vulnerability could be addressed through responding to climate change. Conventionally, it was argued that addressing energy vulnerability meant ensuring that those individuals and households who needed to were enabled to consume *more* energy, and, hence, the potential for reducing demand and mitigating climate change was regarded as negligible. More recently, however, some initiatives have begun to examine how interventions to improve the energy efficiency of dwellings, coupled with the provision of low-carbon energy supplies, could serve not only to address climate change, but also to provide secure forms of energy that entail lower running costs for residents. At the same time, it is recognized that the impacts of climate change may serve to exacerbate the vulnerability of residents to poor housing, particularly in terms of extremes of heat and cold, and that, as climate-change policies begin to take hold at the national level, vulnerable communities could bear the brunt of any price rises. Responding to climate change at the urban and neighbourhood level is then framed as a matter, not of *urban* security per se, but of securing the livelihoods of particular urban groups and of reducing their vulnerability.

One example is that of an initiative being undertaken in Melbourne, as a partnership between the Moreland Energy Foundation and the Brotherhood of St Laurence, in which addressing climate change is seen both in terms of enhancing the resilience of the poorer communities in society and in terms of building an alternative economy (Brotherhood of St Laurence 2012). As part of their Moreland Solar City project, funded by the Australian federal government's Solar City programme, these organizations have established a *Warm Homes, Cool Homes* programme, which provides residents with a free advice and installation service designed to increase the energy efficiency of homes and, hence, reduce the vulnerability to both cold and hot weather that the city experiences (Moreland Solar City 2012). This will be delivered through a 'community enterprise' model that 'will provide training and employment, and deliver practical energy efficiency services to low income households' (Moreland Solar City 2012).

There is also evidence that experiments that seek to address the challenges of poverty and economic development alongside climate change are emerging in cities in countries in the Global South. One of the most well-known examples is the Kuyasa project in the Khayelitsha area of Cape Town. Led by the NGO SouthSouthNorth, the project involved providing an energy upgrade to low-income housing, including retrofitting ceilings, energy-efficient light bulbs and solar hot-water heating, which together reduced energy use in households (hence, yielding carbon savings) and energy poverty, providing direct financial benefits. Unlike other projects of this nature that are taking place in cities in the Global South, the Kuyasa initiative is particularly innovative because of its use of the CDM as one source of finance. As a result, the project has created local employment opportunities, as well as providing direct financial and carbon savings (see SouthSouthNorth 2011). Many initiatives that take a 'pro-poor' approach to climate-change adaptation being developed variously by community-based groups and partnerships between municipalities and other agencies could also be considered under this form of alternative experimentation (see Chapter 6).

Climate resilience and the making of community

For a second set of alternative climate-change experiments, the concept of climate resilience is also critical. Rather than focusing on the resilience of individuals or households, however, these alternative experiments focus on fostering different forms of *community* based resilience and response to climate change. At least in part, this is a response driven by a concern for some of the more radical and disruptive effects that climate change might have, including the disruption and collapse of existing forms of economic and social organization. Initiated in Totnes in the UK and now to be found in over 400 cities in thirty-four countries, focused on the UK, North America, Asia and Australia, the Transition Towns movement is one such alternative (North 2010, Smith 2010):

> A Transition Initiative is a community . . . working together to look Peak Oil and Climate Change squarely in the eye and address this BIG question: for all those aspects of life that this community needs in order to sustain

itself and thrive, how do we significantly increase resilience (to mitigate the effects of Peak Oil) and drastically reduce carbon emissions (to mitigate the effects of Climate Change)?

(Project Dirt 2012)

The notion of transition behind this movement is one in which communities self-organize to address both the 'peak-oil' challenge and climate change through the development of a more localized economy, through, for example, the production of local sources of food and energy. In common with the discourse of 'secure urbanism', regarded by Hodson and Marvin (2010) as characteristic of contemporary urban climate governance, for Transition Towns, the promotion of self-sufficiency is regarded as a means of achieving resilience. However, rather than being based on an abstract notion of the city or of the securitization of dominant political and economic interests, central to this vision is the engagement of a notion of community and of resilience in the face of the likely collapse of existing institutions. As Transition Town Brixton, London, explains:

> Transition Town Brixton is a community-led initiative that seeks to raise awareness locally of Climate Change and Peak Oil. TTB proposes that it is better to design that change, reduce impacts and make it beneficial than be surprised by it. We will vision a better low energy/carbon future for Brixton. We will design a Brixton Energy Descent Plan. And then we will make it happen.
>
> (Transition Town Brixton 2011a)

Transition Town initiatives encompass a wide range of actions and activities, and, although they have predominantly focused on small towns, the ideas of the transition movement are increasingly being adopted in large cities such as London, Birmingham and Nottingham through the creation of smaller-scale urban communities. In Brixton, sixteen different groups have been formed, addressing issues including education, arts and culture, recycling and reusing materials, energy conservation, local food production and the development of a local currency, the Brixton pound (see the city case study on London). Although framed in terms of the transition of the community, the actual work of Transition Towns initiatives spans interventions that aim to enhance resilience at the household and community level. For example, a 'draught-busting' initiative seeks to engage with householders in draught proofing their homes in order to save energy, carbon and money, and also provides loans of smart meters so that householders can assess the effectiveness of their own efforts to reduce energy use (Transition Town Brixton 2011b). Like other Transition Town initiatives, within Brixton there is a strong focus on the development of alternative sources of food within the community, including the development of community gardens, bee-keeping, seed sharing and planting 'edible' trees. Transition Towns do not, therefore, only provide an alternative set of possible interventions and actions in response to the insecurity of climate change, but offer different visions for what sustainable and resilient urban futures might look like. Although some might suggest that such visions are hopelessly romantic, they serve as a reminder that the political consequences of addressing climate change in the city are not always tied into the continued domination of current patterns of political economy.

CITY CASE STUDY

Climate experiments and alternatives in London

London's emissions of GHGs are substantial and similar to those of some European countries, such as 'Greece or Portugal' (London Climate Change Agency 2007: 1). Since 2000, the formation of the GLA, active political leadership and the support of key economic and community groups in the city have placed climate change high on the political agenda. In 2007, the first London Climate Change Action Plan adopted the ambitious target of stabilising 'CO_2 emissions in 2025 at 60 per cent below 1990 levels, with steady progress towards this over the next 20 years' (Greater London Authority 2007: 19). Alongside such strategic visions and policy programmes, responding to climate change in London has involved a range of different forms of experiment and alternative, driven both by government actors and by grass-roots organizations.

One approach to cutting carbon dioxide in London has been the designation by the GLA of ten low-carbon zones across the city, with the aim of bringing about a 20.12 per cent reduction in carbon emissions by 2012, in time for the London Olympics. Local authorities were invited to bid for funding for an area of up to 1,000 residential, commercial and public-sector buildings. One of these low-carbon zones is Brixton, an inner-city area of south London and part of the London Borough of Lambeth. Brixton Low Carbon Zone (LCZ) was launched in March 2010 and contains around 3,500 properties, including ten high-rise and thirty-six low-rise blocks, street properties (social and private housing) and commercial and public-sector buildings. Much of the zone falls within one of the most deprived wards in England, with high unemployment, high levels of fuel poverty and the majority of properties consisting of social housing.

The aim of the LCZ is to help residents and businesses within the zone to cut their carbon emissions, reduce waste and save energy, supported by a project officer, waste officer and community-engagement officer. Key projects developed by the LCZ include Green Doctors and Community Draughtbusters – mechanisms for installing energy-efficiency measures in homes and talking to households about energy use – and Green Community Champions. Parallel to this, the LCZ has supported local projects emerging from community strengths, such as food growing and community gardening. The existence of LCZ has also been used to lever investment for capital improvements, with funding successfully gained for a community energy-saving programme on the Loughborough estate.

Although headline targets focus on carbon reduction, in its implementation the LCZ has been less focused on climate change, and more focused on building capacity within the local community as a mechanism for addressing carbon targets in the long term. In this way, what began as a 'government-led' experiment is dependent for its long-term success on the creation of new forms of alliance within and between communities.

An alternative, grassroots approach to addressing the challenges raised by climate change is Transition Town Brixton (TTB). Part of the wider Transition Town movement, TTB was the first urban Transition Town when it was launched in 2007. It arose from key individuals – already active in Lambeth – being enthused by the transition model that was being implemented in Totnes and seeking to apply the same model in Brixton. TTB seeks to raise local awareness of climate change and peak oil and plan a transition to a better, low-energy future. In line with the Transition Town movement, the emphasis is on 're-localizing' and on building local community resilience in order to achieve this.

TTB focuses on the entire area of Brixton, and the number of active volunteers ranges from 100 to 200, although, as TTB is organized into a 'hub and spoke model', with a central working group alongside a number of different thematic working groups, not everyone is associated with all elements of TTB. Reflecting these organizational arrangements, the projects undertaken under the umbrella of TTB are broad and include the ongoing development of the Brixton pound (an alternative local currency), Brixton Energy Group and Invisible Food, among a variety of others.

TTB focuses on incremental change over time and sees part of its role as being an early adopter – both in the local community but also in relation to other transition groups across London. This emphasis on time and the importance of the visionary nature of the transition process is significant and means the focus is on helping people to feel better about where they are and what they can do in order to be able to control how they go forward, rather than simply being about climate change.

Several interconnections between Brixton LCZ and TTB, and the benefits of these interconnections, are apparent. There is a longstanding relationship between Lambeth council and TTB, and building on this relationship and other community engagement initiatives in Brixton was a key part of Lambeth's bid for the LCZ. As a result, TTB are represented on the steering group for the LCZ and have been key partners in delivering projects under the remit of the LCZ, such as the community draughtbusters. Furthermore, the LCZ, partly because of its dedicated funding stream, has allowed TTB to progress projects that were proving difficult, such as discussions with the Housing Association about solar panels on buildings.

Neither the LCZ nor TTB focuses solely on carbon reduction, and instead they view climate change as a more holistic agenda, with equal – if not more – importance placed on building community capacity and local resilience. In this respect, one important difference is the notion of community within such initiatives. The LCZ model of community is based on a specific geographical area and number of buildings, which does not immediately translate into a single, readily identifiable community, whereas TTB has a much wider remit and a more organic notion of community. Despite this, one challenge for both

Figure 7.4 Community gardens in Brixton, London
Source: Sara Fuller

initiatives is reaching out beyond the usual suspects. The LCZ has a community engagement officer whose role is to facilitate community projects and connect people, but challenges remain in engaging so-called 'hard-to-reach' groups. Conversely, where TTB might be expected to be better able to reach such groups owing to its grass-roots nature, it lacks capacity for outreach work and has relied on the LCZ to reach a wider and more diverse set of participants.

In summary, although the LCZ has the advantage of a dedicated funding stream to support projects, designing and implementing projects to achieve specific carbon targets within a two-year time frame is challenging. In contrast, TTB is able to benefit from a more fluid notion of community, is focused more on visioning and has longevity on its side, but sometimes struggles to find resources to support its work. This suggests that, in the long run, one key role of government-led climate-change experiments might be to facilitate the development and growth of grass-roots initiatives.

Sara Fuller, Department of Geography, University of Durham, UK

Transition Towns are one area in which experimentation is taking place that not only seeks to respond to climate change, but also serves to foster new forms of community. A second set of alternative experiments that act in this manner are community-based urban tree-planting programmes. Rather than focusing on individual benefits, such programmes tend to stress the collective and civic benefits that tree planting in the city has, and, albeit often more implicitly than explicitly, to emphasize both the potential adaptation and mitigation gains. One such scheme is the MillionTreesNYC campaign, heralded as a key part of the PlaNYC strategy for sustainability and carried out in partnership between the NYC Department of Parks and Recreation and the New York Restoration Project, 'a civil society organization focused on enhancing underused green spaces throughout the city' (Fisher *et al.* 2011: 8). In her work on this project, Dana Fisher demonstrates how the scheme encapsulates the idea of 'voluntary stewardship' – participants are asked, through both workplace schemes and as individual householders, to volunteer to plant trees, which are then cared for through local stewardship groups. Researching the MillionTreesNYC campaign, Fisher and her colleagues found that, compared with the population of the city, volunteers were more often female, white and highly educated, reflecting general trends in volunteers across the USA, and that they became involved with the campaign through personal and organizational networks. Rather than being a sign of the growing individualization that commentators frequently suggest is taking place in contemporary Western societies, Fisher and colleagues (2011: 27) argue that participants were 'digging together'; the MillionTreesNYC can be regarded as a collective endeavour, through which organizational and social networks are built and reinforced.

Radical sustainability?

While Transition Towns and other forms of community-based response to climate change advocate alternative discourses and practices that take account of some of the radical disruption that the conjunction of climate change and peak oil could produce, for some they fall short of the sort of radical response that seeks to transform the existing social, political and economic structures through which climate change is being produced and that structure vulnerability. As discussed in Chapter 6, transformative responses to climate change entail moving beyond existing structures of decision-making and modes of organization, to foster new regimes and practices through which climate change is addressed.

Examples of experiments that wholly pursue this form of climate politics are rare, although it may be possible to identify parts of projects that contain this sort of radical edge, including in the Transition Towns initiatives discussed above. One example of where these principles of a more radical, transformative approach to climate change have been pursued is in protest actions and social-movement organizations, such as is found in the Climate Camp and the Occupy movement. In Toronto, climate activists who had participated in action at the 2009 Copenhagen Summit and the 2010 World People's Conference on Climate Change and the Rights of Mother Earth in Cochabamba, Bolivia, organized the Toronto People's Assembly on Climate Justice (2011). The Toronto People's Assembly aimed to:

> create a space where we can work together to share experience, knowledge, and resources in order to build a local response to a global crisis. The Assembly hopes to work towards this objective through channels of collective dialogue and community empowerment.
> (Toronto People's Assembly on Climate Justice 2012)

Rather than seeking to act directly to reduce GHG emissions or develop resilience, actions such as the Toronto People's Assembly seek to bring together like-minded groups in an effort to reframe the policy debate, and to show the potential of alternative ways of considering the climate-change problem – here as one of social justice. Carbon Rationing Action Groups (CRAGs) have a similar ethos of seeking to develop a more radical approach to sustainability that recognizes the need for those that consume the most to reduce GHG emissions first and furthest. Through a network of people concerned with climate change, individuals come together in CRAGs to set a collective target to live within a 'carbon budget' and pay penalties for exceeding this level of consumption (see Community Pathways 2012). Although the overall impact of CRAGs, confined to small groups and a handful of cities, may be small, the idea demonstrates that alternative forms of urban living are possible. Low-impact developments, often collective forms of building alternative dwellings that have a minimal carbon footprint over their lifecycle, are also forms of experiment that seek to demonstrate the viability of radically alternative technologies and practices (Pickerill 2010). In the UK, the Centre for Alternative Technology has a long history of advancing innovation in this field, and the Low Carbon Trust has recently been established to promote alternative approaches to building and construction. As not-for-profit organizations, both have sought to pioneer new techniques and to show the potential for different kinds of technology. For example, the Low Carbon Trust developed the Earthship Brighton building as a community centre and demonstration of the viability of alternative building materials, such as rammed-earth tyres, renewable energy and rainwater-harvesting systems (Low Carbon Trust 2012). As with many climate-change experiments, here, it is not so much the technology itself that is new, but its demonstration within a particular urban context and its use as a means of achieving particular social objectives – in this case, affordable, self-build housing with a low environmental impact.

Sustaining alternatives

Climate-change experiments have provided different forms of alternative to mainstream, liberal environmental responses to climate change. In some cases, these have been built on an accommodation between mainstream discourses and the concerns to harness the potential for climate change to address challenges of urban poverty and social justice. For others, climate change can become a means by which resilience and community can be fostered, harnessing the potential for some degree of transition from existing forms of political economy to alternative economies and forms of social organization. For a third set of experiments, radical alternatives to mainstream accounts of the climate problem and how it should be solved are required – whether this takes the form of protest and dissent, the articulation of responsibilities among the affluent to reduce their consumption, or advocacy of alternative technologies.

Sustaining or scaling up such alternatives is a significant challenge. As Smith (2007: 436) argues, 'green niches', most notably grass-roots innovations, are 'constructed in opposition to incumbent regimes. They are informed, initiated and designed in response to sustainability problems perceived in the regime'. There are, however, various examples of how the principles, technologies and practices embedded in such forms of alternative experimentation have entered into more mainstream responses – technologies such as microgeneration schemes and rainwater harvesting, which were once regarded as marginal to the building industry and are increasingly included in urban development projects, mandated by local governments, and taken up by individual householders and businesses. Notions of transition, resilience and climate justice make their presence felt on urban, national and international agendas, often supported by evidence pointing to the ways in which they are being tried and tested in these sorts of experiment. Grass-roots innovations, and other alternatives, can therefore move from the margins to the mainstream (Seyfang 2009).

However, there remain important distinctions between mainstream responses to climate change and the kinds of alternative documented here. First, rather than being integrated into existing forms of economy, these alternative experiments offer models of low-carbon living, where forms of social and technical innovation are put to work to create new forms of economic and community relations. Second, these alternative experiments explicitly recognize that resource security is an essentially contested and unequal concept, with the result that vulnerability and resilience are highly differentiated within the city. These alternative forms of innovation,

> point to a fractured landscape of urban climate change responses, where strange bedfellows (e.g. international carbon finance and low income households in South Africa) are conjoined in developing new discourses of security and resilience, and where the potential for contestation and conflict is ever present.
>
> (Bulkeley and Betsill 2011)

In turn, this suggests that urban climate politics cannot automatically be regarded as 'a politics reduced to the administration and management of processes whose parameters are defined by consensual socio-scientific knowledges' (Swyngedouw 2009: 602), but, rather, that, 'conflict, albeit sometimes latent and worked through everyday practices of resistance, contestation and the formation of the alternative, is emerging over what climate change should mean and for whom, and of the consequences for the future of cities' (Bulkeley and Betsill 2011).

Conclusions

This chapter has suggested that, alongside the concerted efforts by municipalities to plan and implement mitigation and adaptation strategies, a wealth of responses to climate change are emerging in cities on a project-by-project basis. Rather than viewing these initiatives as one-off, isolated examples, this chapter has argued that they need to be seen as an integral part of the way in which cities are responding to climate change. These climate-change experiments are emerging in cities across different regions of the world and, although predominantly led by municipal actors, are also being undertaken by a range of business and civil-society organizations. The resulting landscape is not one in which a smooth layer of climate governance has settled over existing urban actors, processes and governance structures, but rather one that resembles a patchwork comprised of the interweaving of policies and projects, plans and initiatives, a diverse array of actors and agendas, and significant absences.

Examining experiments in practice, the chapter argued that they can be found in at least four different domains – policy innovation, eco-city developments, novel technologies and the transformation of social practices. Urban climate-change experiments provide a means by which the logics of municipal voluntarism and strategic urbanism that have underpinned mainstream urban responses to climate change can be demonstrated and given vitality, serving to reinforce these ways of responding to the climate challenge in cities. Rather than being separate from mainstream urban responses to climate change, experimentation can be seen as part of this response – scaling up experiments will therefore require, not the transfer of learning from individual projects into broader-scale policy processes, but the recognition that existing approaches to governing climate change have produced this piecemeal response, and that moving beyond this will require more fundamental change in this governance process. At the same time, there is evidence that experiments provide a means by which alternative discourses of climate change are given voice – as those seeking to advance agendas of addressing urban poverty and social justice seek to engage experiments as a means of illustrating the potential of climate change to achieve these ends, the mobilization of resilience and community around the climate-change issue forges new forms of intervention in cities, and radical approaches to sustainability are articulated. Given that such experiments are often developed precisely in opposition to mainstream approaches to urban climate governance, the challenge here is perhaps one of how their radical potential can both be maintained and used to inform the transformation of existing regimes and practices.

Discussion points

- What might concepts of sociotechnical regimes and their transition offer for understanding the ways in which climate-change experiments are emerging in cities? What are the advantages and limitations of these approaches?
- Why might non-state actors, including multinational corporations, non-governmental organizations, universities and charitable foundations regard cities as a place in which they can experiment with responses to climate change? Choose some examples to compare and contrast.
- Examine websites, blogs, YouTube videos, the activities of a local Transition Town group and consider to what extent the discourses of transition, resilience and community articulated in Transition Towns represent a challenge to orthodox approaches to urban climate-change governance.

Further reading and resources

There are several different sources of literature that explore the concept of governance and various forms of social and technical experiment. Some good starting points include:

Bulkeley, H. and Castán Broto, V. (2012a) Government by experiment? Global cities and the governing of climate change, revised for *Transactions of the Institute of British Geographers* (awaiting acceptance).

Evans, J. P. (2011) Resilience, ecology and adaptation in the experimental city, *Transactions of the Institute of British Geographers*, 36: 223–37.

Farrelly, M. and Brown, R. (2011) Rethinking urban water management: experimentation as a way forward? *Global Environmental Change*, 21(2): 721–32.

Hoffmann, M. J. (2011) *Climate Governance at the Crossroads: Experimenting with a Global Response after Kyoto*. Oxford University Press, Oxford.

Raven, R. (2007) Niche accumulation and hybridisation strategies in transition processes towards a sustainable energy system: an assessment of differences and pitfalls, *Energy Policy*, 35: 2390–400.

Seyfang, G. and Smith, A. (2007) Grassroots innovations for sustainable development: towards a new research and policy agenda, *Environmental Politics*, 16: 584–603.

Information about different kinds of experiment and alternative taking place in cities can be found by looking at the websites of different state and non-state actors in cities. Just a few examples are provided here for illustration:

- Masdar City: www.masdar.ae/en/Menu/index.aspx?MenuID=48&CatID=27&mnu=Cat
- OneMillionTreesNYC: www.milliontreesnyc.org/html/home/home.shtml
- Transition Towns: www.transitionnetwork.org/ (see also YouTube for short films by different Transition Town groups)
- WWF Hong Kong: www.wwf.org.hk/en/whatwedo/footprint/climate/.

Notes

1. This research project was sponsored by the ESRC (Award Number: RES-066-27-0002), and the team was led by Harriet Bulkeley. The survey was organized and conducted by Dr. Vanesa Castan Broto (UCL), with assistance from Andrea Armstrong and Anne Maassen.
2. The example is drawn from the *Carbon, Control and Comfort* (an EPSRC–E.ON Strategic Partnership Consortium EP/G000395/1) project and, in particular, work with Karen Bickerstaff and Emma Hinton on low-carbon energy initiatives in London.

8 Conclusions

Climate change is not simply happening to cities, as a suite of environmental processes and events that cities need to endure and overcome. Rather, climate change is actively being *produced* through the urban condition. The multiple forms of social, economic and environmental vulnerability that structure urban societies will mediate the impacts of climate change, shaping the ways in which climate change is experienced, the consequences of particular climate-related events and the possibilities for adaptation. At the same time, the production of GHG emissions through urban economies, patterns of urban consumption and the practices of urban residents will have a significant effect on the global atmosphere and on the extent to which dangerous climate change is realized.

As we have seen throughout this book, cities themselves are not unchanged by the climate-change phenomenon. The urban experience is punctuated by climate-related risks, from severe storms to flood events, heatwaves to new forms of disease, in turn shaping millions of daily decisions about how to cope with adverse conditions and adapt to future risk, investment decisions about where to locate businesses, and policy choices about the provision of infrastructure, the protection of assets and urban development. Cities have also become a site through which multiple actors have sought to mitigate climate change. New forms of mobility, energy provision, architecture, urban regeneration scheme, community action and everyday behaviour have been created in cities in response to the challenge of reducing GHG emissions. In this manner, climate change is serving to reconfigure the city, producing new forms of urbanism that jostle alongside existing urban structures, political economies and cultures.

This book has illustrated the interdependence and mutual constitution of climate change and cities. However, it is vital to remember that this is not a totalizing phenomenon, one in which the production and experience of climate change are the same everywhere and urban responses are universal. As this book has shown, the diversity of urban conditions and livelihoods and the multiple aspects of the climate-change problem are creating a highly uneven landscape. For some urban residents, climate change is a daily reality with which they have to cope in extremely adverse circumstances. For others, climate change means that their local supermarket has a new photovoltaic panel. Within the same city, and across different urban contexts, the effects of climate change can be profoundly different and the opportunities and challenges of responding to this phenomenon diverse. Although the notion of a 'global' environmental problem such as climate change may appear to provide the promise of a collective response, in which social, political and economic differences are put to the side in order to take on a common cause, the reality in cities across the world is that climate change is currently serving to exacerbate, not reduce, inequalities.

This concluding chapter summarizes the main topics and debates discussed in the book and considers their implications for the future of urban responses to climate change. In the first section, the connections between vulnerability and adaptation, risk and mitigation, experiments and alternatives are considered, and the key issues are summarized. In the second section, the extent to which current approaches are capable of addressing the challenge of climate justice is examined, and the implications for urban futures are considered.

Changing climate, changing cities?

Throughout this book, the relation between climate change and the city has been considered through an examination of six key topics: vulnerability, the production of GHG emissions, mitigation, adaptation, experimentation and alternatives. The ways in which urban actors are responding to these challenges have been framed in terms of their coping strategies, different means and modes of governing, and the multiple forms of intervention that are emerging through experimentation and the forging of alternative responses. This section draws together these topics in order to consider the main debates about how cities can address vulnerability, adapt to climate change and reduce their GHG contribution through mitigation, and the roles of experimentation and alternatives in these responses.

Addressing vulnerability: urban adaptation to climate change

Much of the debate about how cities will be affected by climate change has focused on the impacts that they will experience – from droughts to floods and from sea-level rise to changing temperature regimes. As Chapter 2 explained, a great deal of attention and effort is being expended in order to determine the sorts of impact that cities are likely to experience, to provide the evidence base upon which to build assessments of vulnerability and adaptation strategies. Such evidence is often a critical requirement in terms of attracting political attention and the resources to place climate change on urban agendas. However, it can also be costly in terms of time and money, and is often constrained by the limits of scientific models of climate impacts at the local scale and the availability of local data. Such forms of impact assessment are therefore unlikely to take place in all but the largest cities, or in those in which the threat of climate change is already perceived as significant.

Concerns have also been raised that a focus on the impacts of climate change has been to the detriment of understanding *vulnerability*. That is, that, although impact studies can provide important information about the sorts of climate change that might be experienced, they might not sufficiently engage with understanding the ways in which existing urban contexts, conditions and processes shape vulnerability. As a result, studies that have mapped vulnerability to climate impacts have tended to focus on broad-scale risks to urban *locations*, such as those that are exposed to coastal and river flooding. Such assessments may provide an overview of the sorts of scale of economic assets and populations at risk, but they are limited in their ability to show how and why different urban communities may be vulnerable to such risks, and, hence, offer only a partial basis upon which to build an adaptation response.

Alternative assessments of climate impacts and vulnerability have started from a different position – seeking to examine the ways in which the physical and economic development of urban *places* serves to structure exposure to risk and vulnerability. These studies point to the ways in which urban vulnerability is mediated through the built environment and infrastructure networks that provide cities with services of shelter, energy provision, health, water, sanitation and so on, and that it is most significant in the poorer and informal parts of cities. Across these landscapes, vulnerability is also differentiated within and between urban *communities*, according to differences between households and individuals in terms of the risks that they may face, their ability to cope and their adaptive capacity. Urban poverty is critical in shaping these patterns of urban vulnerability, but so too are other attributes, including gender, age, employment and social networks. Climate vulnerability in the city can therefore be regarded as fashioned through the interaction between climate impacts, the physical and economic production of urban places, and social variations in terms of coping and adaptive capacity.

These different interpretations of climate vulnerability matter for how *adaptation* is conceived and takes place. For most of the world's urban residents, responding to climate change has been a matter of coping – of using existing resources to seek to reduce the risks to which they are exposed. Moving from coping to adaptation, where 'an actor is able to reflect upon and enact change in those practices and underlying institutions that generate root and proximate causes of risk, frame capacity to cope and further rounds of adaptation to climate change' (Pelling 2011a: 21), requires significant changes in both approach and resources. The evidence suggests that such forms of purposive adaptation are only now beginning to emerge in cities in the form of strategies, plans and measures. Although it may be the case that climate-change considerations are being worked into a range of decisions, such as building standards and where urban development should be located, we have limited understanding of the way in which this is being undertaken and the consequences that it is having. Focusing on the policy response to adaptation, Chapter 6 showed how both expert-led and community-based forms of adaptation are emerging. However, for the most part, adaptation remains focused on what Pelling (2011a) has termed *resilience*, working within existing social, economic and political structures in order to improve the capacity to cope with risk, rather than entailing approaches that move towards a *transition* or *transformation* in institutions and practices in order fundamentally to alter how adaptation takes place. This suggests that, for all the recognition that vulnerability is a deep-seated and structural issue within cities, for the most part efforts to adapt to climate change have, as yet, failed to respond to this challenge.

Uneven geographies of GHG emissions and the challenge of mitigation

In contrast to adaptation, the *mitigation* of climate change has received sustained attention at the urban level over the past 20 years. During this time, significant effort has been put into understanding the contribution of cities to global GHG emissions. Various organizations have developed different approaches, methodologies, tools and metrics through which to measure the GHG emissions that are produced in cities. Although policymakers and practioners now routinely suggests that over 70 per cent of energy-related carbon dioxide emissions are produced by cities, this figure, and more importantly what it means, has been hotly debated. Conventional GHG-emissions accounting allocates responsibility to the *producers* of emissions, rather than *consumers* to whom products and services are delivered. This, in turn, means that it is those cities in which the production of GHG emissions is most significant that are seen to be contributing most to global climate change. In some contexts, this is uncontroversial. For example, the energy used to heat buildings and the fuel used to drive cars are regarded as both produced and consumed in the same location. However, when it comes to considering the embodied energy in goods that are produced in one city and exported to another, such as electronic goods or food, the GHG emissions are allocated to the producer rather than the consumer. This, critics suggest, is unjust, for it leads to a situation of blaming those cities, which are often in the developing world and already suffer the negative consequences of the environmental and labour conditions of manufacturing industries, for the problem of climate change, which is actually driven by increasing levels of consumption elsewhere. Furthermore, by continuing to focus on GHG emissions at the point of production, urban consumers gain little understanding about how their daily lives are intricately connected to the production of GHG emissions and to climate change. New forms of accounting, such as carbon footprinting, are being used to try to convey this information, but, for the most part, these GHG emissions remain hidden from view.

This view of the urban climate-mitigation challenge, as one of reducing those GHG emissions produced in the city, has been pervasive. Policy approaches have followed a relatively standard model in which emissions are measured, targets are set, and plans are devised. Although this has been relatively successful within the well-resourced and navigable territory of municipal GHG emissions within economically developed cities, as the imperative for, and interest in, mitigation has spread beyond these confines, this approach has become more challenging. On the one hand, as mitigation has come on to the agenda in a more diverse range of urban political and economic contexts, the resources and capacity to undertake such an approach have often been found wanting. On the other hand, measuring, monitoring, setting targets and implementing policy within the urban community writ large have proven to be complex and often contested tasks. Nonetheless, as Chapter 5 demonstrated, there is significant evidence that climate-change mitigation is becoming part of how cities undertake urban development, design and regenerate the built environment, and reconfigure urban infrastructure networks and services. Although generic trends are visible in terms of a focus on the enabling and self-governing modes of governance and a tendency to focus on the energy sector and energy efficiency, the extent and nature of climate-change mitigation in particular urban contexts have been shaped by the highly differentiated levels of institutional capacity that exist between cities, as well as by the political and sociotechnical challenges that have been encountered.

As a result of these challenges, although individual cities, and the transnational municipal networks of which they are often part, can point to successes in terms of GHG emissions reduced and additional benefits, particularly in terms of addressing energy security and economic efficiency and meeting other societal needs, the gap between the targets set and the achievements made by and large remains. Over the past two decades, the role of cities in producing GHG emissions and in mitigating climate change has moved from the margins to the mainstream of many policy agendas at the urban, national and international levels. However, despite the many thousands of cities that have established targets, developed plans and taken action, the reality is that, for most cities and most urban residents across the world, mitigating climate change remains either a distant concern or a luxury that cannot be afforded.

Creating a transition? Experimentation and alternative climate-change responses

The analysis in this book suggests that, at the urban level, we can see growing recognition of the challenges of climate vulnerability and GHG emissions in cities, the development of concerted mitigation efforts, and nascent efforts to support adaptation and resilience. However, it is also apparent that the responses to these challenges have not emerged in a coherent or universal manner, with evidence gathered, targets set and policies implemented. Instead, this has been mediated through existing political concerns, institutional capacities and sociotechnical networks, and has been shaped by the changing nature of the (local) state and engagement of a host of non-state actors with the climate-change agenda. Rather than finding a smooth urban landscape, where addressing climate change is readily slotted into public and private agendas, the reality has been more complex, highly uneven and often contested. One result of these processes has been that the purposive governing of climate change has not taken place purely through policy, strategies and plans, but has, instead, also been conducted through the development of specific projects and initiatives. At the same time, the ubiquity of climate change has meant that it has become attached to different kinds of urban intervention, site and actor. The result is a patchwork of climate-change responses in the city, being conducted by different actors and operating across different scales and networks, often for different purposes and towards different ends.

It could be argued that this uneven landscape marks a failure of climate governance – a failure to create the coherence, vision and scale of change required adequately to respond to the climate challenge. What is required is the integration of these parts into a larger whole, the making of one urban response. Indeed, many urban climate-change strategies and plans bear these hallmarks, as they variously collect and stich together projects and initiatives taking place across the city. Equally, that a thousand climate-change flowers are blooming could be the cause for celebration, acknowledging that their presence offers a much needed supplement to the efforts of urban climate-change policy and that they might provide an opportunity for learning and improving the policy response. Despite the different attitudes to climate-change experiments that these views express, neither acknowledges that such interventions and initiatives are a fundamental part of the urban response to climate change. As James Evans has argued, 'climate experiments are where governance is located; they represent the practical dimension of adaptation [and mitigation] – what happens in practice, "on the ground", when policy-makers, researchers, businesses and communities are charged with finding new paths' (Evans 2011: 225). In this sense, experimentation is not an *option* to be dismissed or embraced, but rather has become a fundamental part of the urban response to climate change. This matters, as Evans (2011: 233), goes on to argue, for, 'if sustainability comes down to letting 1,000 experimental flowers bloom, then it matters who gets to experiment, and how'. As Chapter 7 explained, across the domains of policy innovation, eco-city development, novel technologies and the transformation of social practices, experimentation is dominated by efforts to pursue municipal voluntarism and strategic urbanism, tied up in logics of efficiency, security and carbon control and underpinned by an ideology of environmental liberalism. Nonetheless, experimentation does make space for alternatives, pursuing social and environmental justice, novel forms of transition and resilience, and radical interpretations of sustainability. Although the integration of such principles and practices into the mainstream is far from assured, there is some evidence that such alternative forms of experimentation can create different ways of articulating and responding to the urban climate-change problem.

Urban futures: towards climate justice?

What might the city of the future look like, be like, feel like? The question of urban futures has been one that has exercised scholars and students since the formation of the first cities. As set out in the Introduction to this book, climate change has served to reopen and reconfigure some of these debates. In the popular imagination, and the foundation of many of the disciplines upon which urban studies are built, nature is something that lies *outside* the city (Bulkeley and Betsill 2003; Owens 1992). Over the past two decades, however, there has been a growing recognition of the ways in which cities profoundly shape the opportunities and limits for sustainable development, and of how what is 'natural' is produced through urban as well as rural landscapes. This conceptual accommodation between nature and the city has enabled us to examine climate change, not as something that is happening to, on or around cities, but rather as constituted through and with contemporary urbanism. Urban futures, therefore, cannot be divorced from climate futures. As illustrated in the Introduction, these futures are articulated in terms of various utopian ideals of the designed eco-city or of the self-organizing, climate-resilient community, or in the dystopian images of the city at peril from the changing climate or serving as the engine driving the coming eco-disaster.

These urban fictions, like most, contain a grain of truth. Each of these utopian and dystopian notions of the climate-changed city can be found in some shape or form across the highly differentiated landscape that is the urban response to climate change. This book has shown examples of how eco-design, community-based adaptation, resilience and numerous other utopian principles are being taken up, reformed and put to work in pursuit of adaptation and mitigation. Equally, it has shown how the impacts of climate change pose very real risks to urban communities, and how continuing processes of urban production and consumption are, albeit in a highly uneven manner, continuing to drive the global atmosphere towards dangerous climate change. Among this uneven and fragmented picture, there is a very real sense of the need to get at the realities of the response to climate change: just what have cities achieved? What must they do? How can they do it? The evidence is scarce, with the challenges of data collection, monitoring and measurement discussed above, together with concerns about damaging the fragile political will to address climate change at the urban level, limiting our knowledge of what is being achieved in concrete terms. Nonetheless, transnational municipal networks, such as the Climate Alliance, ICLEI and C40, together with individual cities, can point to some significant achievements in terms of the evidence gathered, targets set, emissions reduced and schemes in place to enhance adaptive capacity.

Is this enough? The first, and most straightforward, answer is clearly not. As suggested above, climate change has yet to be a concern for most cities and remains distant from the daily lives and struggles of most urban residents. While urban vulnerability persists and GHG emissions continue to rise, this is indeed an 'urgent agenda' (World Bank 2010). A second response is more positive. In comparison with the slow progress being made by the international community to address the climate-change problem, that cities are acting *at all* could be considered something of a miracle. Rather than seeking to measure the effectiveness of urban responses to climate change, in terms of GHG emissions reduced or resilience improved, we could instead consider the broader effects of these responses in terms of raising the profile of how mitigation and adaptation can be achieved, of showing how these issues are fundamentally linked to questions of urban economic and social development, and of placing the city on the international agenda.

A third response to the question of whether the urban response to climate change has been sufficient is more complex and perhaps most important. What is clear from the analysis in this book is that the challenges of climate change and the responses to them are highly uneven, both within and between cities. Rather than corresponding to one or other utopian or dystopian ideal, the urban landscape comprises different elements of these discourses and their iteration and interaction. As such, there is no such thing as *the* urban climate-change problem, but rather a series of urban processes and practices through which climate change comes to matter in particular places and for particular communities. However, despite acknowledging this difference, for the most part the urban governance of climate change – those purposive and strategic interventions being carried out by political and economic elites in various cities across the world – tends to assume a universal challenge to be addressed. Although particular vulnerable groups may be identified for special treatment, climate impacts are considered as a challenge for the whole city to address. At the same time, urban climate-change strategies identify targets for mitigation that are levelled at the scale of the city, without acknowledging the different contributions that different kinds of resident, business, visitor and so on may be making. Despite the need to use the civic and collective identity that the notion of the city brings in order to advance climate change as a political issue, this kind of universalism raises very significant challenges in terms of whether urban responses can adequately address questions of climate justice. Rather than being based on the attribution of responsibility or rights equally across a city, principles of justice suggest that what is required is recognition of how the costs being borne, the opportunities afforded and the capacity to participate in decision-making are differentiated across urban communities, and measures should be taken to ensure that this is taken into consideration in the design and implementation of climate policy. Developing an adequate response to climate change in the city, therefore, is not only about whether urban vulnerability is reduced, GHG emissions are curtailed and the wider importance of the urban response is realized, but it is also fundamentally about recognizing difference and marshalling this towards a collective response to the climate-change challenge.

Bibliography

Adelekan, I. O. (2010) Vulnerability of poor urban coastal communities to flooding in Lagos, Nigeria, *Environment and Urbanization*, 22(2): 433–50.

Adger, W. N., Arnell, N. A. and Tompkins, E. L. (2005) Successful adaptation to climate change across scales, *Global Environmental Change*, 15(2): 77–86.

Adger, W. N., Dessai, S., Goulden, M., Hulme, M., Lorenzoni, I., Nelson, D. R., Naess, L. O., Wolf, J. and Wreford, A. (2009) Are there social limits to adaptation to climate change? *Climatic Change*, 93: 335–54.

Ahammad, R. (2011) Constraints of pro-poor climate change adaptation in Chittagong City, *Environment and Urbanization*, 23(2): 503–15.

Akinbami, J. F. and Lawal, A. (2009) *Opportunities and challenges to electrical energy conservation and CO_2 emissions reduction in Nigeria's building sector*. Paper prepared for the Fifth Urban Research Symposium, Cities and Climate Change: Responding to an Urgent Agenda, 28–30 June, Marseille, France.

Alam, M. and Golam Rabbani, M. D. (2007) Vulnerabilities and responses to climate change for Dhaka, *Environment and Urbanization*, 19(1): 81–97.

Alber, G. and Kern, K. (2008) Governing climate change in cities: modes of urban climate governance in multi-level systems. *Proceedings of the OECD Conference on Competitive Cities and Climate Change*. OECD, Paris.

Allen, J. (2004) The whereabouts of power: politics, government and space, *Geografiska Annaler*, 86B(1): 19–32.

Allman, L., Fleming, P. and Wallace, A. (2004) The progress of English and Welsh local authorities in addressing climate change, *Local Environment*, 9(3): 271–83.

Anguelovski, I. and Carmin, J. (2011) Something borrowed, everything new: innovation and institutionalization in urban climate governance, *Current Opinion in Environmental Sustainability*, 3: 169–175.

Arup (2011a) Infographic: how are cities tackling climate change? Online: www.arup.com/Homepage_Cities_Climate_Change.aspx#!lb: /Homepage_Cities_Climate_Change/Infographic.aspx (accessed January 2012).

Arup (2011b) *Climate Action in Mega Cities: C40 cities baseline and opportunities*, Version 1.0, June, ARUP.

Aulisi, A., Larsen, J., Pershing, J. and Posner, P. (2007) *Climate Policy in the State Laboratory: How States Influence Federal Regulation and the Implications for*

Climate Change Policy in the United States. World Resources Institute, Washington DC.

Aylett, A. (2010) Municipal bureaucracies and integrated urban transitions to a low carbon future, in Bulkeley, H., Castán Broto, V., Hodson, M. and Marvin, S. (Eds) *Cities and Low Carbon Transition*, Routledge, Abingdon and NewYork, pp. 142–58.

Aylett, A. (2011) Changing Perceptions of Climate Mitigation Among Competing Priorities: The Case of Durban, South Africa, in *Cities and Climate Change: Global Report on Human Settlements 2011*, UN-HABITAT.

Bai, X. (2007) Integrating global environmental concerns into urban management: the scale and readiness arguments, *Journal of Industrial Ecology*, 11(2): 15–29.

Bartlett, S. (2008) Climate change and urban children: impacts and implications for adaptation in low-and middle-income countries, *Environment and Urbanization*, 20(2): 501–19.

Bernstein, S. (2001) *The Compromise of Liberal Environmentalism*. Columbia University Press, New York.

Berrang-Ford, L., Ford, J. D. and Paterson, J. (2011) Are we adapting to climate change? *Global Environmental Change*, 21: 25–33.

Betsill, M. M. and Bulkeley, H. (2006) Cities and the multilevel governance of global climate change, *Global Governance*, 12(2): 141–59.

Betsill, M. and Bulkeley, H. (2007) Looking back and thinking ahead: a decade of cities and climate change research, *Local Environment: The International Journal of Justice and Sustainability*, 12(5): 447–56.

Bicknell, J., Dodman, D. and Satterthwaite, D. (Eds) (2009) *Adapting Cities to Climate Change: Understanding and Addressing the Development Challenges*. Earthscan, London.

Bioregional (2009) Capital consumption: the transition to sustainable consumption and production in London, November 2009, Bioregional and London Sustainable Development Commission. Online: www.bioregional.com/files/publications/capital-consumption.pdf (accessed January 2012).

Birkmann, J., Garschargen, M., Kraas, F. and Quang, N. (2011) Adaptive urban governance: new challenges for the second generation of urban adaptation strategies to climate change, *Sustainability Science*, 5: 185–206.

Birkmann, J. and von Teichman, K. (2010) Integrating disaster risk reduction and climate change adaptation: key challenges – scales, knowledge, and norms, *Sustainability Science*, 5: 171–84.

Blake, J. (1999) Overcoming the 'value–action gap' in environmental policy: tensions between national policy and local experience, *Local Environment*, 4: 257–78.

Brody, S. D., Zahran, S., Vedlitz, A. and Grover, H. (2008) Examining the relationship between physical vulnerability and public perceptions of global climate change in the United States, *Environment and Behavior*, 40(1): 72–95.

Brotherhood of St Laurence (2012) Climate change. Online: www.bsl.org.au//Research-and-Publications/Research-and-Policy-Centre/Climate-change

Bulkeley, H. (2000) Down to earth: local government and greenhouse policy in Australia, *Australian Geographer*, 31(3): 289–308.

Bulkeley, H. (2001) No regrets? Economy and environment in Australia's domestic climate change policy process, *Global Environmental Change*, 11: 155–69.

Bulkeley, H. (2009) Planning and governance of climate change, in Davoudi, S., Crawford, J. and Mehmood, A. (Eds) *Planning for Climate Change Strategies for Mitigation and Adaptation for Spatial Planners*, Earthscan, London.

Bulkeley, H. (2010) Cities and the governing of climate change, *Annual Review of Environment and Resources*, 35: 229–53.

Bulkeley, H. (2012, forthcoming) Climate change and urban governance: a new politics? In Lockie, S., Sonnenfeld, D. and Fisher, D. (Eds) *International Handbook of Social and Environmental Change*, Routledge, London.

Bulkeley, H. and Betsill, M. M. (2003) *Cities and Climate Change: Urban Sustainability and Global Environmental Governance*, Routledge, London.

Bulkeley, H. and Betsill, M. M. (2011) Revisiting the urban politics of climate change, submitted to *Environmental Politics*, in revised form, June 2012.

Bulkeley, H. and Castán Broto, V. (2012a) Government by experiment? Global cities and the governing of climate change, revised for *Transactions of the Institute of British Geographers* (awaiting acceptance).

Bulkeley, H. and Castán Broto, V. (2012b) Urban experiments and the governance of climate change: towards Zero Carbon Development in Bangalore, *Contemporary Social Science*, accepted subject to revision.

Bulkeley, H. and Kern, K. (2006) Local government and climate change governance in the UK and Germany, *Urban Studies*, 43: 2237–59.

Bulkeley, H. and Newell, P. (2010) *Governing Climate Change*, Routledge, Abingdon.

Bulkeley, H. and Schroeder, H. (2009) *Governing Climate Change Post-2012: The Role of Global Cities – Melbourne*. Tyndall Centre for Climate Change Research Working Paper 138.

Bulkeley, H., Schroeder, H., Janda, K., Zhao, J., Armstrong, A., Chu, S. Y. and Ghosh, S. (2009) *Cities and Climate Change: The Role of Institutions, Governance and Urban Planning*. Report for the World Bank Urban Research Symposium: Cities and Climate Change.

Bulkeley, H., Watson, M. and Hudson, R. (2007) Modes of governing municipal waste, *Environment and Planning A*, 39(11): 2733–53.

C40 (2011a) Fact Sheet: Why Cities? Online: http://c40citieslive.squarespace.com/storage/FACTper cent20SHEETper cent20Whyper cent20Cities.pdf (accessed October 2011).

C40 (2011b) C40 Releases Groundbreaking Research on the Importance and Impact of Cities on Climate Change. Online: c40citieslive.squarespace.com/storage/C40%20Research%20Press%20Release.pdf (accessed January 2012).

C40 Cities (2011c) C40 São Paulo Summit. Retrieved 26 January 2012 from www.c40saopaulosummit.com/site/conteudo/index.php?in_secao=26

C40 Cities (2011d) C40 São Paulo Summit letter to Rio + 20, United Nations Conference on Sustainable Development. Retrieved 26 January 2012 from www.c40saopaulosummit.com/site/conteudo/index.php?in_secao=37&ib_home=1

C40 Cities (2012) C40 Voices: Adalberto Maluf reports on recent advances in São Paulo's bus transit strategy. Online: http://live.c40cities.org/blog/2012/1/12/c40-voices-adalberto-maluf-reports-on-recent-advances-in-sao.html

Carbon Disclosure Project (2008) *CDP cities 2011: Global report on C40 cities*, CDP/KPMG. Online: https://www.cdproject.net/CDPResults/65_329_216_CDP-CitiesReport.pdf.

CFU (2010a) *Carbon Finance at the World Bank*. World Bank, Washington DC.

CFU (2010b) *A City-wide Approach to Carbon Finance*. World Bank, Washington DC.

Carter, J. G. (2011) Climate change adaptation in European cities, *Current Opinion in Environmental Sustainability*, 3: 193–8.

Cartwright, A. (2008) Final report: sea-level rise adaptation and risk mitigation measures for the city of Cape Town, prepared by by Anton Cartwright (SEI Cape Town) in collaboration with Professor G. Brundrit and Lucinda Fairhurst, July 2008. Online: www.capetown.gov.za/en/EnvironmentalResourceManagement/publications/Documents/Phase%204%20-%20SLRRA%20Adaptation+Risk%20Mitigation%20Measures.pdf.

Castán Broto, V. and Bulkeley. H. (2012) A survey of urban climate change experiments in 100 global cities, submitted to *Global Environmental Change*.

Castleton, H. F., Stovin, V., Beck, S. B. M. and Davison, J. B. (2010) Green roofs: building energy savings and the potential for retrofit, *Energy and Buildings*, 42: 1582–91.

CDP (2008) Carbon Disclosure Project Cities Pilot Project 2008, Report by the CDP for ICLEI US, available online: www.cdproject.net/CDPResults/65_329_216_CDP-CitiesReport.pdf (accessed February 2012).

Chaterjee, M. (2010) Slum dwellers response to flooding events in the megacities of India, *Mitigation and Adaptation Strategies for Global Change*, 15: 337–53.

City of Cape Town Environmental Resources Management Department (2009) *Enviroworks: Special Edition Energy and Climate Change*. Online: www.capetown.gov.za/en/EnvironmentalResourceManagement/publications/Documents/Enviroworks_Dec09.pdf (accessed January 2012).

City of London (2010) *Rising to the Challenge – The City of London Climate Change Adaptation Strategy*. First published May 2007; revised and updated January 2010. City of London. Online: www.cityoflondon.gov.uk/services/environment-and-planning/sustainability/Documents/pdfs/SUS_AdaptationStrategyfinal_2010update.pdf.

City of Melbourne (2008) Zero Net Emissions by 2020 Update, Arup Pty Ltd for the City of Melbourne, Melbourne, City of Melbourne.

City of Melbourne (2012) CH2 – Water conservation. Online: http://www.melbourne.vic.gov.au/Sustainability/CH2/aboutch2/Pages/WaterConservation.aspx.

City of Philadelphia (2007) Local action plan for climate change, Sustainability Working Group, April 2007. Online: www.dvgbc.org/green_resources/library/city-philadelphia-local-action-plan-climate-change, last accessed 30 June 2011.

City of Philadelphia (2009) Greenworks Philadelphia, City of Philadelphia, Mayor's Office of Sustainability. Online: www.phila.gov/green/greenworks/2009-greenworks-report.html, last accessed 30 June 2011.

City of Philadelphia (2011) *Greenworks Philadelphia 2011 Progress Report*, City of Philadelphia, Mayor's Office of Sustainability.

Coafee, J. and Healy, P. (2003) My voice my place: tracking transformations in urban governance, *Urban Studies*, 40(10): 1979–99.

Collier, U. (1997) Local authorities and climate protection in the European Union: putting subsidiarity into practice? *Local Environment*, 2: 39–57.

Commission on Climate Change and Development (2009) *Closing the Gaps: Disaster Risk Reduction and Adaptation to Climate Change in Developing Countries*, Commission on Climate Change and Development, Stockholm, Sweden.

Community Pathways (2012) Carbon rationing action group (CRAG) or energy saving club, available online: http://www.communitypathways.org.uk/approach/439/full (accessed September 2012).

Copenhagen Climate Communiqué (2009) Copenhagen Climate Communiqué, Copenhagen 2009. Online: www.kk.dk/Nyheder/2009/December/~/media/B5A397DC695C409983462723E31C995E.ashx, last accessed January 2012.

Corburn, J. (2009) Cities, climate change and urban heat island mitigation: localising global environmental science, *Urban Studies*, 46(2): 413–27.

Corfee-Morlot, J., Cochran, I., Hallegate, S. and Teasdale, P. J. (2011) Multilevel risk governance and urban adaptation policy, *Climatic Change*, 104: 169–97.

Coutard, O. and Rutherford, J. (2010) The rise of post-network cities in Europe? Recombining infrastructural, ecological and urban transformation in low carbon transitions, in Bulkeley, H., Castán Broto, V., Hodson, M. and Marvin, S. (Eds) *Cities and Low Carbon Transition*, Routledge, Abingdon, pp. 107–25.

Covenant of Mayors (2011a) About the covenant. Online: www.eumayors.eu/about/covenant-of-mayors_en.html (accessed January 2012).

Covenant of Mayors (2011b) Welcome. Online: www.eumayors.eu/home_en.htm (accessed January 2012).

Davoudi, S., Crawford, J. and Mehmood, A. (Eds) (2009) *Planning for Climate Change: Strategies for Mitigation and Adaptation for Spatial Planners*. Earthscan, UK and USA.

Department for Communities and Local Government (2007) *Improving the Flood Performance of New Buildings: Flood Resilient Construction*, Department for Communities and Local Government, London.

Dhakal, S. (2011) *Urban energy transitions in Chinese cities*, in Bulkeley, H., Castrán Broto, V., Hodson, M., Marvin, S. (Eds) Cities and Low Carbon Transitions, Routledge, pp. 73–87.

Dockside Green (2012) A better approach. Online: www.docksidegreen.com/Sustainability/Ecology.aspx (accessed January 2012).

Dodman, D. (2009) Blaming cities for climate change? An analysis of urban greenhouse gas emissions inventories. *Environment and Urbanization*, 21(1): 185–201.

Dodman, D., Kibona, E. and Kiluma, L. (2011) Tomorrow is too late: responding to social and climate vulnerability in Dar es Salaam, Tanzania, case study prepared for *Cities and Climate Change: Global Report on Human Settlements 2011*. Online: www.unhabitat.org/grhs/2011 (accessed January 2012).

Douglas, I., Alam, K., Maghenda, M., McDonnell, Y., McLean, L. and Campbell, J. (2008) Unjust waters: climate change, flooding and the urban poor in Africa, *Environment and Urbanization*, 20: 187–205.

Dubeux, C. and La Rovere, E. (2011) The contribution of urban areas to climate change: the case study of São Paulo, Brazil, case study prepared for *Cities and Climate Change: Global Report on Human Settlements 2011*.

Eakin, H. and Lemos, M. C. (2006) Adaptation and the state: Latin America and the challenge of capacity-building under globalization, *Global Environmental Change*, 16(1): 7–18.

Eakin, H., Lerner, A. M. and Murtinho, F. (2010) Adaptive capacity in evolving peri-urban spaces: responses to flood risk in the Upper Lerma River Valley, Mexico, *Global Environmental Change*, 20: 14–22.

Environment Agency (2011) Thames Estuary 2100. Online: www.environment-agency.gov.uk/homeandleisure/floods/104695.aspx (accessed December 2011).

European Environment Agency (2011) Greenhouse gas emission trends and projections in Europe 2011 – Tracking progress towards Kyoto and 2020 targets. Online: www.eea.europa.eu/publications/ghg-trends-and-projections-2011 (accessed January 2012).

Evans, J. P. (2011) Resilience, ecology and adaptation in the experimental city, *Transactions of the Institute of British Geographers*, 36: 223–37.

Farrelly, M. and Brown, R. (2011) Rethinking urban water management: experimentation as a way forward? *Global Environmental Change*, 21(2): 721–32.

Fisher, D., Connolly, J., Svendsen, E. and Campbell, L. (2011) *Digging Together: Why People Volunteer to Help Plant One Million Trees in New York City*. Environmental Stewardship Project at the Center for Society and Environment of the University of Maryland White Paper 1.

Foresight (2008) *Powering our Lives: Sustainable Energy Management and the Built Environment*. Final Project Report, The Government Office for Science, London.

Friends of the Earth Hong Kong (2011) Friends of the Earth Power Smart Contest 2011. Online: www.foe.org.hk/powersmart/2011/e_index.html (accessed January 2012).

Geels, F. W. (2002) Technological transitions as evolutionary reconfiguration processes: a multi-level perspective and a case study, *Research Policy*, 31(8–9): 1257–74.

Geels, F. W. and Kemp, R. (2007) Dynamics in socio-technical systems: typology of change processes and contrasting case studies, *Technology in Society*, 29: 441–55.

Global Cool Cities Alliance (2012) What is global cool cities alliance? Online: www.globalcoolcities.org/ (accessed January 2012).

Gore, C. and Robinson, P. (2009) Local government response to climate change: our last, best hope? in Selin, H. and VanDeveer, S. D. (Eds) *Changing Climates in*

North American Politics: Institutions, Policymaking and Multilevel Governance, MIT Press, Cambridge, MA, pp. 138–58.

Gore, C., Robinson. P. and Stren, R. (2009) *Governance and Climate Change: Assessing and Learning from Canadian Cities*. Fifth Urban Research Symposium Cities and Climate Change: Responding to an Urgent Agenda, Marseille.

Graham, S. and Marvin, S. (2001) *Splintering Urbanism: Networked Infrastructures, Technological Mobilities and the Urban Condition*. Routledge, London.

Granberg, M. and Elander, I. (2007) Local governance and climate change: reflections on the Swedish experience, *Local Environment*, 12: 537–48.

Greater London Authority (2007) *Action Today to Protect Tomorrow: the Mayor's Climate Change Action Plan*, Greater London Authority, London, February.

Greater London Authority (2012) London heat map: welcome. Online: www.londonheatmap.org.uk/Content/home.aspx

Gurran, N., Hamin, E. and Norman B. (2008) *Planning for Climate Change: Leading Practice Principles and Models for Sea Change Communities in Coastal Australia*, Report no. 3 for the National Sea Change Taskforce, July 2008.

Gustavsson, E., Elander, I. and Lundmark, M. (2009) Multilevel governance, networking cities, and the geography of climate-change mitigation: two Swedish examples, *Environment and Planning C: Government and Policy*, 27: 59–74.

Hammer, S. (2009) *Capacity to act: the critical determinant of local energy planning and program implementation*. Paper presented at the Fifth Urban Research Symposium, Cities and Climate Change: Responding to an Urgent Agenda, Marseille.

Handmer, J. W. and Dovers, S. R. (1996) A typology of resilience: rethinking institutions for sustainable development, *Organization and Environment*, 9(4): 482–511.

Hanson, S., Nichols, R., Ranger, N., Hallegate, S., Corfee-Morlot, J. C., Herweijer, C. and Chateau, J. (2011) A global ranking of port cities with high exposure to climate extremes, *Climatic Change*, 104: 89–111.

Hardoy, J. and Pandiella, G. (2009) Urban poverty and vulnerability to climate change in Latin America, *Environment and Urbanization*, 21: 203–24.

Hardoy, J. and Romero-Lankao, P. (2011) Latin American cities and climate change: challenges and options to mitigation and adaptation responses, *Current Opinion in Environmental Sustainability*, 3: 1–6.

Hargreaves, T., Burgess, J. and Nye, M. (2010) Making energy visible: a qualitative field study of how householders interact with feedback from smart energy monitors, *Energy Policy*, 38(10): 6111–19.

Harries, T. and Penning-Rowsell, E. (2011) Victim pressure, institutional inertia and climate change adaptation, *Global Environmental Change*, 21: 188–97.

Hegger, D. L. T., Van Vliet, J. and Van Vliet, B. J. M. (2007) Niche management and its contribution to regime change: the case of innovation in sanitation, *Technology Analysis & Strategic Management*, 19(6): 729–46.

Hillmar-Pegram, K. C., Howe, P. D., Greenberg, H. and Yarnal, B. (2011) A geographic approach to facilitating local climate governance: from emissions inventories to mitigation planning, *Applied Geography*, 34: 76–85.

Hobson, K. and Neimeyer, S. (2011) Public responses to climate change: the role of deliberation in building, *Global Environmental Change*, 21: 957–71.

Hodson, M. and Marvin, S. (2009) 'Urban ecological security': a new urban paradigm? *International Journal of Urban and Regional Research*, 33(1): 193–215.

Hodson, M. and Marvin, S. (2010) *World Cities and Climate Change: Producing Urban Ecological Security*. Open University Press, Milton Keynes.

Hoffman, M. (2009) *Experimenting with Climate Governance*. 2009 Conference on the Human Dimensions of Global Environmental Change – Earth System Governance: People, Places and the Planet, Amsterdam.

Hoffmann, M. J. (2011) *Climate Governance at the Crossroads: Experimenting with a Global Response after Kyoto*. Oxford University Press, Oxford.

Holgate, C. (2007) Factors and actors in climate change mitigation: a tale of two South African cities. *Local Environment*, 12(5): 471–84.

Hommels, A. (2005). Studying Obduracy in the City: Toward a Productive Fusion between Technology Studies and Urban Studies, *Science, Technology & Human Values*, 30: 323–51.

Hoornweg, D., Sugar, L. and Gomez, C. L. T. (2011) Cities and greenhouse gas emissions: moving forward, *Environment and Urbanization*, 23(1): 207–27.

Hulme, M. (2009) *Why We Disagree About Climate Change*. Cambridge University Press, Cambridge.

Hunt, A. and Watkiss, P. (2011) Climate change impacts and adaptation in cities: a review of the literature, *Climatic Change*, 104: 13–49.

Huq, S., Kovats, S., Reid, H. and Satterthwaite, D. (2007) Reducing risks to cities from disasters and climate change, *Environment and Urbanization*, 19(3): 3–15.

ICLEI (1997) *Local Government Implementation of Climate Protection: Report to the United Nations*. International Council for Local Environmental Initiatives, Toronto.

ICLEI (2006) *ICLEI International Progress Report – Cities for Climate Protection*. ICLEI, Oakland.

ICLEI Australia (2008) *Local Government Action on Climate Change: Measures Evaluation Report 2008*. Australian Government Department of Environment, Water, Heritage and the Arts and ICLEI, Melbourne, Australia. Online: http://www.iclei.org/fileadmin/user_upload/documents/Global/Progams/CCP/CCP_Reports/ICLEI_CCP_Australia_2008.pdf

ICLEI Australia (2009) CCP Australia December 2008–January 2009. Online: www.iclei.org/index.php?id=9264#c34751 (accessed January 2012).

ICLEI (2009) *International Local Government GHG Emissions Analysis Protocol Version 1.0 (October 2009)*. Online: www.iclei.org/index.php?id=ghgprotocol

ICLEI (2011) *Global Standard on Cities Greenhouse Gas Emissions – C40 and ICLEI MoU*. Online: www.iclei.org/index.php?id=1487&tx_ttnewsper cent5Btt_newsper cent5D=4643&tx_ttnewsper cent5BbackPidper cent5D=983&cHash=712a8184bb

ICLEI (2012) *The Five Milestone Process*. Online: www.iclei.org/index.php?id=810 (accessed January 2012).

IPCC (2007a) Synthesis report: summary for policy-makers 1. Online: www.ipcc.ch/publications_and_data/ar4/syr/en/spms1.html (accessed January 2012).
IPCC (2007b) *Climate Change 2007: Working Group III: Mitigation of Climate Change – Glossary E-I*. Online: www.ipcc.ch/publications_and_data/ar4/wg3/en/annex1-ensglossary-e-i.html (accessed January 2012).
IPCC (2007c) Synthesis report: summary for policy-makers 2. Online: www.ipcc.ch/publications_and_data/ar4/syr/en/spms2.html (accessed January 2012).
IPCC (2007d) Synthesis report: summary for policy-makers 3. Online: www.ipcc.ch/publications_and_data/ar4/syr/en/spms3.html (accessed January 2012).
IPCC (2007e) *Climate Change 2007: Working Group II: Impacts, Adaptation and Vulnerability – Glossary A-D*. Online: www.ipcc.ch/publications_and_data/ar4/wg2/en/annexessglossary-a-d.html (accessed January 2012).
IPCC (2007f) *Climate Change 2007: Working Group II: Impacts, Adaptation and Vulnerability – Glossary P-Z*. Online: www.ipcc.ch/publications_and_data/ar4/wg2/en/annexessglossary-p-z.html
International Energy Agency (2008) *World Energy Outlook 2008*. International Energy Agency, Paris.
International Energy Agency (2009) *Cities, Towns and Renewable Energy: Yes in My Front Yard*. IEA, Paris.
Jackson, B., Lee-Woolf, C., Higginson, F., Wallace, J. and Agathou, N. (2009) *Strategies for reducing the climate impacts of red meat/dairy consumption in the UK*. WWF/Imperial College London, London. Online: http://assets.wwf.org.uk/downloads/imperialwwf_report.pdf (accessed January 2012).
Jessup, B. and Mercer, D. (2001) Energy policy in Australia: a comparison of environmental considerations in New South Wales and Victoria, *Australian Geographer*, 32(1): 7–28.
Jollands, N. (2008) *Cities and Energy: A Discussion Paper*. OECD International Conference on Competitive Cities and Climate Change. OECD, Milan.
Jones, L. and Boyd, E. (2011) Exploring social barriers to adaptation: insights from Western Nepal, *Global Environmental Change*, 21: 1262–74.
Joss, S. (2010) Ecocities: a global survey, *WIT Transactions on Ecology and The Environment*, 129: 239–50.
Karl, T. R., Melillo, J. M. and Peterson, T. C. (Eds) (2009) *Global Climate Change Impacts in the United States*. Cambridge University Press, New York.
Kennedy, C., Pinsetl, S. and Bunje, P. (2010) The study of urban metabolism and its applications to urban planning and design, *Environmental Pollution*, 159: 1965–73.
Kern, K. and Bulkeley, H. (2009) Cities, Europeanization and multi-level governance: governing climate change through transnational municipal networks, *Journal of Common Market Studies*, 47: 309–32.
Kiithia, J. (2011) Climate change risk responses in East African cities: need, barriers and opportunities, *Current Opinion in Environmental Sustainability*, 3: 1–5.
Kingdon, R. W. (1984) *Agenda, Alternatives and Public Policies*. Longman, London.

Kirshen, P., Ruth, M. and Anderson, W. (2008) Interdependencies of urban climate change impacts and adaptation strategies: a case study of Metropolitan Boston USA, *Climatic Change*, 86: 105–22.

Koehn, P. H. (2008) Underneath Kyoto: emerging subnational government initiatives and incipient issue-bundling opportunities in China and the United States, *Global Environmental Politics*, 8(1): 53–77.

Krause, R. M. (2011) Symbolic or substantive policy? Measuring the extent of local commitment to climate protection, *Environment and Planning C: Government and Policy*, 29(1): 46–62.

Lambright, W. H., Chagnon, S. A. and Harvey, L. D. D. (1996) Urban reactions to the global warming issue: agenda setting in Toronto and Chicago, *Climatic Change*, 34: 463–78.

Lasco, R., Lebel, L., Sari, A., Mitra, A. P., Tri, N. H. (Eds) (2007) *Integrating Carbon Management Into Development Strategies of Cities – Establishing a Network of Case Studies of Urbanisation in Asia Pacific*. Final Report for the APN project 2004–07-CMY-Lasco.

Laukkonen, J., Blanco, P. K., Lenhart, J., Keiner, M., Cavric, B. and Kinuthia-Njenga, C. (2009) Combining climate change adaptation and mitigation measures at the local level, *Habitat International*, 33: 287–92.

Leach, R. and Percy-Smith, J. (2001) *Local governance in Britain*. Palgrave Macmillan, London.

Lebel, L., Huaisai, D., Totrakool, D., Manuta, J. and Garden, P. (2007) A carbon's eye view of urbanization in Chiang Mai: improving local air quality and global climate protection, in Lasco, R., Lebel, L., Sari, A., Mitra, A. P., Tri, N. H. *et al.* (Eds) *Integrating Carbon Management Into Development Strategies of Cities – Establishing a Network of Case Studies of Urbanisation in Asia Pacific*. Final Report for the APN project 2004–07-CMY-Lasco, pp. 98–124.

Leichenko, R. (2011) Climate change and urban resilience, *Current Opinion in Environmental Sustainability*, 3: 164–8.

Levine, S., Ludi, E. and Jones, L. (2011) *Rethinking Support for Adaptive Capacity to Climate Change*. Overseas Development Institute, London. Online: http://policy-practice.oxfam.org.uk/publications/rethinking-support-for-adaptive-capacity-to-climate-change-198311 (accessed January 2012).

Li, J. (2011) Decoupling urban transport from GHG emissions in Indian cities – a critical review and perspectives, *Energy Policy*, 39: 3503–14.

de Loë, R., Kreutzwiser, R. and Moraru, L. (2001) Adaptation options for the near term: climate change and the Canadian water sector, *Global Environmental Change*, 11: 231–45.

LogiCity (2012) Introduction. Online: www.logicity.co.uk/ (accessed January 2012).

London Climate Change Agency (2007) *Moving London Towards a Sustainable Low-Carbon City: An Implementation Strategy*. London Climate Change Agency, London, June.

London Development Agency (2012) *Low Carbon Economy*. Online: www.lda.gov.uk/our-work/low-carbon-future/low-carbon-economy/index.aspx (accessed February 2012).

López-Marrero, T. and Tschakert, P. (2011) From theory to practice: building more resilient communities in flood-prone areas, *Environment and Urbanization*, 23(1): 229–49.

Lovelock, J. (2009) *The Vanishing Face of Gaia: A Final Warning*. Allen Lane, London.

Low Carbon Trust (2012) Low Carbon Trust. Online: www.lowcarbon.co.uk/home

Liu, J. and Deng, X. (2011) Impacts and mitigation on climate change in Chinese cities, *Current Opinion in Environmental Sustainability*, 3: 1–5.

Lynas, M. (2004) *Six Degrees: Our Future on a Hotter Planet*. Harper Perennial, London.

McEwan, I. (2010) *Solar*. Jonathan Cape, London.

McGranahan, G., Deborah Balk and Bridget Anderson (2007) The rising tide: assessing the risks of climate change and human settlements in the low elevation coastal zone, *Environment and Urbanization*, 19: 17–37.

McKibben, B. (2011) *Eaarth: Making a Life on a Tough New Planet*. St Martin's Griffin, New York.

Manuel-Navarrete, D., Pelling, M. and Redclift, M. (2011) Critical adaptation to hurricanes in the Mexican Caribbean: development visions, governance structures, and coping strategies, *Global Environmental Change*, 21: 249–58.

Measham, T. G., Preston, B. L., Smith, T. F., Brooke, C., Gorddard, R., Withycombe, G. and Morrison, C. (2011) Adapting to climate change through local municipal planning: barriers and challenges, *Mitigation and Adaptation Strategies for Global Change*, 16: 889–909.

Moreland Solar City (2011) Energy Hub. Online: www.morelandsolarcity.org.au/our-programs/energy-hub (accessed January 2012).

Moser, S. C., Kasperson, R. E., Yohe, G. and Agyeman, J. (2008) Adaptation to climate change in the Northeast United States: opportunities, processes and constraints, *Mitigation and Adaptation Strategies for Global Change*, 13: 643–59.

NAGA (2006) Northern Alliance for Greenhouse Action Strategic Plan. NAGA, Melbourne.

NAGA (2008) Towards zero net emissions in the NAGA region. NAGA, Melbourne, December.

Neuhäuser, A. (2010) *KWK Modellstadt Berlin – Energy Efficiency in Energy Consumption*. Online: www.kwk-modellstadt-berlin.de/media/file/95.Achim-Neuh%E4user-Berlin-Energy-Agency.pdf

NCC (2012) City consumption. Online: www.newcastle.nsw.gov.au/environment/climate_cam/climatecam (accessed January 2012).

Newman, P. and Kenworthy, J. (1999) *Sustainability and Cities: Overcoming Automobile Dependence*. Island Press, Washington DC.

Nickson, A. (2011) Cities and climate change: adaptation in London, UK, case study prepared for *Cities and Climate Change, Global Report on Human Settlements*. Online: www.unhabitat.org/downloads/docs/GRHS2011/GRHS2011CaseStudyChapter06London.pdf (accessed January 2012).

North, P. J. (2010) Eco-localisation as a progressive response to peak oil and climate change – a sympathetic critique, *Geoforum*, 41(4): 585–94.
Owens, S. (1992) Energy, environmental sustainability, and land-use planning, in Breheny, M. (Ed.) *Sustainable Development and Urban Form*, Pion, London, pp. 79–105.
Oxford is My World (2012) Welcome to Oxford is My World. Online: www.oxfordismyworld.org/ (accessed January 2012).
Padeco (2010) *Cities and Climate Change Mitigation: Case Study on Tokyo's Emissions Trading System*. World Bank, Washington DC.
Paterson, M. (1996) *Global Warming and Global Politics*. Routledge, London.
Paterson, M. (2007) *Automobile Politics: Ecology and Cultural Political Economy*. Cambridge University Press, Cambridge.
Pearce, F. (2009) Greenwash: the dream of the first eco-city was built on a fiction, *The Guardian*, 23 April.
Pelling, M. (2011a) *Adaptation to Climate Change: From Resilience to Transformation*. Taylor & Francis Books, London.
Pelling, M. (2011b) Urban governance and disaster risk reduction in the Caribbean: the experiences of Oxfam, GB, *Environment and Urbanization*, 23(2): 383–400.
Pew Centre (2011) *Adaptation Planning – What US States and Localities Are Planning*. Pew Centre, Washington DC.
Pickerill, J. (2010) Building liveable cities: low impact developments as low carbon solutions? in Bulkeley, H., Castán Broto, V., Hodson, M. and Marvin, S. *Cities and Low Carbon Transitions*. Routledge, pp. 178–97.
Pitt, D. R. (2010) The impact of internal and external characteristics on the adoption of climate mitigation policies by US municipalities, *Environmental and Planning C: Government and Policy*, 28(5): 851–71.
Prefeitura Do Município De São Paulo (2009) Lei 14.933 de 5 de Junho de 2009. São Paulo.
Prefeitura Do Município De São Paulo (2011) Diretrizes para o Plano de Ação da Cidade de São Paulo para Mitigação e Adaptação às Mudanças Climáticas. São Paulo.
Project Dirt (2012) Transition Towns. Online: http://projectdirt.com/page/transition-towns
Puppim de Oliveira, J. (2009) The implementation of climate change related policies at the subnational level: an analysis of three countries, *Habitat International*, 33(3): 253–9.
Qi, Y., Ma, L., Zhang, H. and Li, H. (2008) Translating a global issue into local priority: China's local government response to climate change, *Journal of Environment and Development*, 17(4): 379–400.
Ranger, N., Hallegatte, S. *et al.* (2011) An assessment of the potential impact of climate change on flood risk in Mumbai, *Climate Change*, 104: 139–67.
Raven, R. (2007) Niche accumulation and hybridization strategies in transition processes towards a sustainable energy system: An assessment of differences and pitfalls, *Energy Policy*, 35: 2390–400.

Rawlani, A. K. and Sovacool, B. K. (2011) Building responsiveness to climate change through community based adaptation in Bangladesh, *Mitigation and Adaptation Strategies for Global Change*, 16: 845–63.

Revi, A. (2008) Climate change risk: an adaptation and mitigation agenda for Indian cities, *Environment and Urbanization*, 20(1): 207–29.

Roberts, D. (2010) Prioritizing climate change adaptation and local level resilience in Durban, South Africa, *Environment and Urbanization*, 22(2): 397–413.

Robinson, P. and Gore, C. (2011) *The spaces in between: a comparative analysis of municipal climate governance and action*. Paper presented at the American Political Science Association, Annual Conference, 1–4 September, Seattle, WA.

Romero-Lankao, P. (2007) How do local governments in Mexico City manage global warming? *Local Environment*, 12(5): 519–35.

Romero-Lankao, P. (2010) Water in Mexico City: what will climate change bring to its history of water-related hazards and vulnerabilities? *Environment and Urbanization*, 22(1): 157–78.

Rosenzweig, C., Major, D. C., Demong, K., Stanton, C., Horton, R. and Stults, M. (2007) Managing climate change risks in New York City's water system: assessment and adaptation planning, *Mitigation and Adaptation Strategies for Global Change*, 12: 1391–409.

Rosenzweig, C., Solecki, B., Hammer, S. and Mehrota, S. (2011) *Climate Change and Cities: First Assessment Report of the Urban Climate Change Research Network*. Cambridge University Press, Cambridge.

Rosenzweig, C. and Solecki, W. (2010) Introduction to 'Climate change adaptation in New York City: building a risk management response'. Ann. New York Acad. Sci., 1196, 13–18.

Rutland, T. and Aylett, A. (2008) The work of policy: actor networks, governmentality, and local action on climate change in Portland, Oregon, *Environment and Planning D: Society and Space*, 26(4): 627–46.

Sanchez-Rodriguez, R., Fragkias, M. and Solecki, W. (2008) *Urban Responses to Climate Change a Focus on the Americas: A Workshop Report*. International Workshop Urban Responses to Climate Change, New York City.

Sari, A. (2007) Carbon and the city: carbon pathways and decarbonization opportunities in Greater Jaxarta, Indonesia, in Lasco, R., Lebel, L., Sari, A., Mitra, P., Tri, N. H., Ling, O. G. and Contreras, A. (Eds) *Integrating Carbon Management Into Development Strategies of Cities: Establishing a Network of Case Studies of Urbanization in Asia Pacific*. Final Report for the APN project 2004–07, CMY, Lasco, pp. 125–51.

Satterthwaite D. (2008a) Cities' contribution to global warming: notes on the allocation of greenhouse gas emissions, *Environment and Urbanization*, 20(2): 539–49.

Satterthwaite, D. (2008b) *Climate Change and Urbanization: Effects and Implications for Urban Governance*. United Nations Expert Group Meeting on Population Distribution, Urbanization, Internal Migration and Development, UN/POP/EGM-URB/2008/16.

Satterthwaite, D. (2011) Editorial: why is community action needed for disaster risk reduction and climate change adaptation? *Environment and Urbanization*, 23(2): 339–49.

Satterthwaite, D., Huq, S., Pelling, M., Reid, H. and Romero-Lankao, P. (2008) *Adapting to Climate Change in Urban Areas: The Possibilities and Constraints in Low- and Middle-Income Nations.* IIED, London.

Schreurs, M. A. (2008) From the bottom up: local and subnational climate change politics, *Journal of Environment and Development*, 17(4): 343–55.

Scott, M., Gupta, S., Jáuregui, E., Nwafor, J., Satterthwaite, D., Wanasinghe, Y. A. D. S.,Wilbanks, T., Yoshino, M., Kelkar, U., Mortsch, L. and Skea, J. (2001) Human settlements, energy, and industry. Climate change 2001: impacts, adaptation, and vulnerability, in McCarthy, J. J., Canziani, O. F., Leary, N. A., Dokken, D. J., White, K. S. (Eds) *Contribution of Working Group II to the Third Assessment Report of the Intergovernmental Panel on Climate Change.* Cambridge University Press, Cambridge, pp. 381–416.

Seelig, S. (2011) A master plan for low carbon and resilient housing: the 35 ha area in Hashtgerd New Town, Iran, *Cities*, 28: 545–56.

Setzer J. (2009) *Subnational and transnational climate change governance: evidence from the state and city of São Paulo, Brazil.* Paper presented at the Fifth World Bank Urban Research Symposium: Cities and Climate Change – Responding to an Urgent Agenda, Marseille.

Seyfang, G. (2009) *The New Economics of Sustainable Consumption: Seeds of Change.* Palgrave Macmillan, Basingstoke.

Seyfang, G. and Smith, A. (2007) Grassroots innovations for sustainable development: Towards a new research and policy agenda, *Environmental Politics*, 16: 584–603.

Sharmer, D. and Tomar, S. (2010) Mainstreaming climate change adaptation in Indian cities, *Environment and Urbanization*, 22(2): 451–65.

Short, J., Dender, K. V. and Crist, P. (2008) Transport policy and climate change, in Sperling, D. and Cannon, J. S. (Eds) *Reducing Climate Impacts in the Transportation Sector.* Springer-Verlag, New York, pp. 35–48.

Shove, E. (2003) *Comfort, Cleanliness and Convenience: The Social Organization of Normality.* Berg, Oxford.

Shove, E. (2010) Beyond the ABC: climate change policy and theories of social change, *Environment and Planning A*, 42(6): 1273–85.

Smith, A. (2007) Translating sustainabilities between green niches and socio-technical regimes, *Technology Analysis and Strategic Management*, 19: 427–50.

Smith, A. (2010) Community-led urban transitions and resilience: performing Transition Towns in a city, in Bulkeley, H., Castán Broto, V., Hodson, M. and Marvin, S. (Eds) *Cities and Low Carbon Transition.* Routledge, Abingdon and New York, pp. 159–77.

Smith, A., Voß, J.-P. and Grin, J. (2010) Innovation studies and sustainability transitions: the allure of the multi-level perspective and its challenges, *Research Policy*, 39: 435–48.

Solar American Cities (2011) *Boston Massachusetts*. Online: www.solaramericacities. energy.gov/Cities.aspx?City=Boston (accessed February 2012).

Solar American Cities (2012) About. Online: http://solaramericacommunities.energy. gov/about/ (accessed February 2012).

Solecki, W., Leichenko, R. and O'Brien, K. (2011) Climate change adaptation strategies and disaster risk reduction in cities: connections, contentions, and synergies, *Current Opinion in Environmental Sustainability*, 3: 135–41.

Source London (2012) *Source London*. Online: https://www.sourcelondon.net/ Scope I.

SouthSouthNorth (2011) *Project Portfolio and Reports*. Online: www.southsouth north.org/ (accessed March 2011).

State of Victoria (2005) *Victorian Greenhouse Strategy Action Plan Update*. The State of Victoria Department of Sustainability and Environment, Melbourne.

Stern, N., Peters, S., Bakhshi, V., Bowen, A., Cameron, C., Catovsky, S., Crane, D., Cruickshank, S., Dietz, S., Edmonson, N., Garbett, S.-L., Hamid, L., Hoffman, G., Ingram, D., Jones, B., Patmore, N., Radcliffe, H., Sathiyarajah, R., Stock, M., Taylor, C., Vernon, T., Wanjie, H. and Zenghelis, D. (2006) *Stern Review: The Economics of Climate Change*. HM Treasury, London.

Sugiyama, N. and Takeuchi, T. (2008) Local policies for climate change in Japan, *Journal of Environment and Development*, 17(4): 424–41.

Swyngedouw, E. (2009) The antinomies of the postpolitical city: in search of a democratic politics of environmental production, *International Journal of Urban and Regional Research*, 33(3): 601–20.

Sydney Olympic Park Authority (2012) Water and catchments. Online: www.sopa. nsw.gov.au/our_park/environment/water (accessed January 2012).

The California Energy Commission (2012) Cool Roofs and Title 24. Online: www.energy.ca.gov/title24/coolroofs/ (accessed January 2012).

The Carbon Trust (2012) Organizational carbon footprints. Online: www.carbontrust. co.uk/cut-carbon-reduce-costs/calculate/carbon-footprinting/pages/organisation-carbon-footprint.aspx (accessed January 2012).

The Climate Group (2011) City partnerships. Online: www.theclimategroup.org/ programs/city-partnerships/ (accessed January 2012).

TMG (2008) Tokyo's proposals on nationwide introduction of cap-and-trade program in Japan, Tokyo Metropolitan Government, Tokyo.

TMG (2009) Tokyo cap-and-trade program: Tokyo ETS. Tokyo Workshop 2009 on Urban Cap & Trade Towards a Low Carbon Metropolis, Tokyo.

Tompkins, E. L., Adger, W. L., Boyd, E., Nicholson-Cole, S., Weatherhead, K. and Arnell, N. (2010) Observed adaptation to climate change: UK evidence of transition to a well-adapting society, *Global Environmental Change*, 20: 627–35.

Toronto Environment Office (2008) *Ahead of the Storm: Preparing Toronto for Climate Change*. City of Toronto. Online: www.toronto.ca/teo/pdf/ahead_of_the_ storm.pdf

Toronto People's Assembly on Climate Justice (2011) People's Assembly on Climate Justice: Earth Day 2011. Online: http://torontopeoplesassembly.wordpress.com/ assemblies/earth-day-2011/(accessed January 2012).

Toronto People's Assembly on Climate Justice (2012) People's Assembly on Climate Justice: about. http://torontopeoplesassembly.wordpress.com/about/ (accessed January 2012).

Transition Town Brixton (2011a) About transition? Online: www.transitiontownbrixton.org/?s=About+transition&submit=Search (accessed March 2011).

Transition Town Brixton (2011b) Buildings and energy. Online: www.transitiontownbrixton.org/category/groups/buildingsandenergy/ (accessed March 2011).

Transition Town Network (2012) What is a transition initiative? Online: www.transitionnetwork.org/support/what-transition-initiative (accessed January 2012).

UNFCCC (n.d.) Convention text. Online: http://unfccc.int/files/essential_background/background_publications_htmlpdf/application/pdf/conveng.pdf (accessed January 2012).

UNFCCC (1992) *United Nations Framework Convention on Climate Change*. United Nations, New York.

UN-Habitat (2008) *State of the World's Cities 2008/2009 – Harmonious Cities*. Earthscan, London and Sterling, VA.

UN-Habitat (2009) *Planning Sustainable Cities: Global Report on Human Settlements 2009*. Earthscan, London.

UN-Habitat (2011) *Global Report on Human Settlements: Cities and Climate Change*. UN-Habitat, Nairobi, Kenya.

Union of Concerned Scientists (2008) *Climate Change in Pennsylvania: Impacts and Solutions for the Keystone State*. Union of Concerned Scientists, Cambridge, MA, 54 pp. Online: www.northeastclimateimpacts.org/ (accessed January 2012).

Urry, J. (2011) *Climate Change and Society*. Polity Press, Cambridge.

van Vliet, B., Chappells, H. and Shove, E. (2005) *Infrastructures of Consumption: Environmental Innovation in the Utility*. Earthscan, London.

While, A., Jonas, A. E. G. and Gibbs, D. (2010) From sustainable development to carbon control: eco-state restructuring and the politics of urban and regional development, *Transactions of the Institute of British Geographers*, 35: 76–93.

Wilbanks, T. J., Romero-Lankao, P., Bao, M., Berkhout, F., Cairncross, S., Ceron, J. P., Kapshe, M., Muir-Wood, R. and Zapata-Marti, R. (2007) Industry, settlement and society, in Parry, M. L., Canziani, O. F., Palutikof, J. P., van der Linden, P. J. and Hansen, C. E. (Eds) *Climate Change 2007: Impacts, Adaptation and Vulnerability*. Contribution of Working Group II to the Fourth Assessment Report of the Intergovernmental Panel on Climate Change, Cambridge University Press, Cambridge, pp. 357–90.

Willis, R., Webb, M. and Wilsdon, J. (2007) *The Disrupters: Lessons for Low-Carbon Innovation From the New Wave of Environmental Pioneers*. National Endowment for Science, Technology and the Arts, London.

Wilson, E. and Piper, J. (2010) *Spatial Planning and Climate Change*. Routledge, Abingdon.

Wolf, J., Adger, W. N., Lorenzoni, I., Abrahamson, V. and Raine, R. (2010) Social capital, individual responses to heat waves and climate change adaptation: an empirical study of two UK cities, *Global Environmental Change*, 20: 44–52.

Wollmann, H. (2004) Local government reforms in Great Britain, Sweden, Germany and France: between multi-function and single purpose authorities, *Local Government Studies*, 30(4): 639–65.

World Bank (2010) *Cities and Climate Change: an Urgent Agenda*. World Bank, Washington DC.

Worldmapper (2012) Greenhouse gases. Online: www.worldmapper.org/display.php?selected=299 (accessed January 2012).

WWF Hong Kong (2012) Climateers. Online: www.climateers.org/eng/contents/ (accessed January 2012).

Yardley, J., Sigal, R. J. and Kenny, G. P. (2011) Heat health planning: the importance of social and community factors, *Global Environmental Change*, 21: 670–9.

Zahran, S., Brody, S. D., Vedlitz, A., Grover, H. and Miller, C. (2008) Vulnerability and capacity: explaining local commitment to climate-change policy, *Environment and Planning C: Government and Policy*, 26: 544–62.

Zimmerman, R. and Faris, C. (2011) Climate change mitigation and adaptation in North American Cities, *Current Opinion in Environmental Sustainability*, 3: 181–7.

Index

Note: *Italic* page numbers indicate figures and tables; **bold** page numbers indicate boxes.

Accra (Ghana) **161**
Action Aid 27
adaptation 15, 37–38, 80, 93, 104, 142–189, 226; and built environment *see* built environment and adaptation; and co-benefits *182*, 185–186; community-based *see* community-based adaptation; and coping *see* coping strategies; deficit 146, 150–151; defined/types of 144–150, **145**; in developed countries 151; drivers/barriers for 179–188, *181–183*; efficacy/limits of 144, 150; further reading/resources on 188–189; future for 187–188; implementation of 159; and informal settlements 160, 168, 170, *181*, 184; and infrastructure *see* infrastructure; institutional factors in 179, 180–184, *181*, 192; and knowledge/data 152–153, 160–161, 164; levels of 148, *149*; low-/middle-income countries and 19, 27, 80–82, 150–151, 152–153, 158, 160–162, 217; mal- *183*, 186–187; and mitigation, compared 185–186, 187; and modes of governance 164, 179, 184; neglect of 143, 163; new models of 190; Philadelphia case study 154–158; policy/planning, development of 150–164; political factors in 179, *182–183*, 184–186; and poverty 155, 158, 159–163, 167; and resilience *see* resilience; and risk/disaster 142, 146, 151, 152–154, 164–167, 179; role of municipal authorities in 151–152; sociotechnical factors in 179, *183*, 186–187, 188; transitional/transformational 15, 148, *149*, 150, 159, 187, 188; and uncertainty 142–143; and urban development 167–171, 179, *181–183*; and urban diversity 150–151; and vulnerability 144, 146, 160, 188, 230–232
adaptive capacity 71, 146, 148, 164, 168, 171, 176, 178, 180, 186, 188, 231–232, 236
additionality 55, 109
Aecom 82
Africa *3*, 27, 34, 80, **161**
age factor 35, 39–42, **39**, *41*, 66
air conditioning 118–119, 121, 123, **124**, *139*
air quality/pollution 32, *33*, *41*, 77, 102, 115, 127
airports 121
Amazon region 75–76
Amman (Jordan) 14, 47, 54–58, *56*; Green Growth Program 55–58, *57–58*, 69
Amsterdam (Netherlands) 75, 76, **131**
anticipatory adaptation **145**, 151
architects 72, 82
Arcosanti (US) 206
Argentina 7, 82, 125
art and climate change 10

Arup 11, 80, *89*, 91, 119, 127
Asia *3*, 7, 29, 34, 76, 78, 103, 132; see also South Asia
Asian Cities Climate Change Resilience Network 80
Asian Disaster Preparedness Centre 167
Australasia *3*, 7
Australia 76, 82, 130, 169, 197; CCP in 78, 85, 86, *88*, 112; climate change experiments in 207, *208*, 216; climate change impacts in *3*; mitigation in 112, 135; see also Melbourne; Newcastle
awareness raising 95, 96, **111**, 150, **172**

Bali Conference (2007) 91
Bangalore (India) 206
Bangladesh 38, **39**, 165–167, 170–171
Baoding (China) 122
Barcelona (Spain) 123, 128
Bartlett, Sheridan 39–40
Beijing (China) 129
Berlin (Germany) 76, 128, 129
Besançon (France) 76
Bhubaneswar (India) **79**
bicycles 127, 128
BioCarbon Fund 54
biodiversity *3*, 36
biofuels 63, 115, **124**, 127–128
Bloomberg, Mayor Michael 111
Bogotá (Colombia) 127, 129
Boston (US) 24, 32, 153
Boulder (US) **131**
Brazil 6, 82, 125, 127, 129; see also São Paulo
Brighton (UK) 224
Britain (UK) 40–42, 52, 65–66, 135, 197, 206; building standards in 171–173; climate change experiments in 197, 206, 211, 216, 224; climate change governance in 74; coastal management in 168–169; GHG emissions targets in 106; planning authorities in 120; Stern Review (2006) 7; Transition Towns in see Transition Towns movement; see also London; Manchester; Nottingham
Brixton LCZ (Low Carbon Zone, London) 218–222, *221*
bromine compounds **2**

BRT (bus rapid transit) systems 114, 115, 127
Brussels (Belgium) 128
Buenos Aires (Argentina) 7, 125
building codes 9, 72, 84, 86, 93, 103, 123, 171–173, 180
building materials 52, 119, 224
built environment and adaptation 171–176, 179, *181–183*, 198; informal practices 175–176; and resilience of buildings 171–173; surface treatments 25, 173; water harvesting/recycling 173, **174**, *175*
built environment and mitigation 107, 122–126; drivers/barriers in 132, *133–134*; energy consumption by 122; and energy demand management 125–126; and energy efficiency 122–123; in Global South 125; municipal powers and 123; retrofitting projects see retrofitting
bus rapid transit (BRT) systems 114, 115, 127
Bush, George W. 78
business sector 1, 6, *41*, 72, **99**, 229

C40 Cities Climate Leadership Group 45–46, 53, 78–80, 82, 84, 192, 194, 236; members *81*; and mitigation 110, 117, 119, 127, 136
Calcutta (India) 7
California (US) 173
Canada 121, 153, 159, 178; see also Toronto
capacity building 55, 76–77, *214*
Cape Town (South Africa) 14, 20, 35, 36–38, *38*, 153, 169; hybridized climate change strategy in 217; mitigation in 118, 122; policy responses in 37–38; vulnerability of 36–37
car ownership/use 65, 108, 119, 120, 127, 132, 233
carbon capture and storage 109
carbon control 83, 130, 191, 198
carbon dioxide (CO_2) **2**, 8, *8*, 46, 108, 109, **203**
Carbon Disclosure Project (CDP) 49, 53, 110, 117

Carbon Finance Capacity Building (CFCB) programme 55
carbon finance/market 47, 52, 54–58, 83, 84; Amman case study 55–58, 69; and co-benefits/additionality 55; criticisms of 54–55; emissions trading schemes **203–205**; sectoral approach in 55; strengths/weaknesses of 57
carbon footprint 51–52, 67, 93, 202, 233
carbon monoxide (CO) 115
Carbon Rationing Action Groups (CRAGs) 223–224
carbon sinks/sequestration 106, 107, 108–109, 121, **172**, 179, 198
Carbon Trust **49**
carbon/emissions intensity 63, 130
CCP (Cities for Climate Change) programme 47, 75, 76, 77–78, 83, 84, 136; in Australia 78, 85, 86, *88*, 112; 'milestones' approach and 110, **111**, 112; in South Asia 78, *79*; Streetlight Management Scheme 129; in US 156
CCX (Chicago Climate Exchange) **203–205**
CDCF (Community Development Carbon Fund) 54, 55
CDM (Clean Development Mechanism) 6, 54, 129, 209
CDP (Carbon Disclosure Project) 49, 53, 110, 117
CEMR (Council of European Municipalities and Regions) 75
Centre for Alternative Technology (UK) 224
CFCB (Carbon Finance Capacity Building) programme 55
Chiang Mai (Thailand) 51–52, 119–120
Chicago (US) 102, 123, 165, 173, 179
Chicago Climate Exchange (CCX) **203–205**
children 5, 39–40, *41*, 155
Chile 82
China 6, 60, *60*, 61–62, 63, 82, 127, 129, 132, 135; low-carbon cities in 121–122; *see also* Shanghai
chlorine compounds **2**
CHP (combined heat and power) 129
churches 72

cities 4–5; central to climate change problem 4, 7–9, *8*; future for 235–237; GHG emissions by 7–8, *8*
Cities and Climate Change Initiative (UN-Habitat) 80
Cities for Climate Change programme *see* CCP
Cities and Climate Change (World Bank 2010) 80
civic capacity 84
civil society *8*, 72, **73**, 116, *133*, 192, 199–200, 209, 226
Clean Development Mechanism (CDM) 6, 54, 129, 209
climate activism 223, 224
Climate Alliance 75–76, 77, 78, 83, 110, 236
climate change: artists' response to 10; controversies surrounding 1; created by cities 7–8, 29–30, 229; evidence for 1–2; as global problem 5–6; growth in science of 5; as location-specific problem 6–7; and responsibilities 14; society's lack of engagement with 1, 2; as urban problem 4, 7–9, *8*, 71
climate change experiments 190–228; eco-cities 15, 121, 192, 206, 226, 235, 236; emissions trading schemes **203–205**; and enabling governing mode 193–194, 209; further reading/resources on 227–228; future for 226, 234–235; in Global South 217; grass-root *see* grass-roots/bottom-up approach; hybridized 215, 216–217; incremental approach 192, 194, **204**, 220; limits/implications of 213–215, *214*; literature on 195–197; living laboratories 196–197; and mainstream responses, compared 225; mapping 197–200; niches and 196, 200, 225; policy innovation 15, 192, 201–206, **203–205**, 226, 235; as reaction to international indecision 195; reconfiguring of social practices 192, 206, 209, 226; resilience and 215, 216, 217–223; sectors/forms/actors in 197–200, *199*, *200*, 226; self-governance 193–194; and social justice 192, 215, 216–217, 223–224, 225, 226,

235; social/technical, focus on 198–199, 206; and sociotechnical regimes 196; sustainability and 215, 223–224, 235; technical innovations 192, 207–212, 226, 235; three criteria for 195–196; Transition Towns 12–13, 82, 215, 217–223
climate change impacts 2–4, **2**, 6, 14, 20–27, 230–231; direct/indirect effects 20, *21*; financial cost of 24–25; in low-/middle-income countries 19, 32–35; positive 20; predicting/assessing *see* predicting climate change; on services to cities 22; variations in 19, 20
Climate Group 78, 102, 201, 223
climate justice 235–237
climate models 23, 25–27; limitations of 27
Climate Partnership (HSBC) 82
Climate Protection Agreement (US) 78
Climate Summit for Mayors (2009) 9–10
ClimateCam 50, **50**, *51*
CLIMB (Climate's Long-term Impacts on Metro Boston) study 32
Clinton Climate Initiative 78, 156, 201
co-benefits 55, 77, *133*, *182*, 185–186
coal 63, 106
coastal areas *3*, 7, *8*; vulnerability of 20, **23**, 28–29, 36–38; *see also* flooding; sea-level rise
coastal erosion *21*, 28–29
coastal management 168–169
Coimbatore (India) **79**
CoM (European Covenant of Mayors) 78, 83
combined heat and power (CHP) 129
commercial buildings 7, 122, 230
communities 12–13, 72, 142, 211; and GHG emissions 47–48; and risks/vulnerabilities 28, 36, **39**, 169, 231
Community Development Carbon Fund (CDCF) 54, 55
community-based adaptation 142, 152, 159–163, 165–167, 187; Accra case study **161**; Durban case study **172**; experiments in 217–222, 224; partnerships and 162–163, 165; resilience and 159, **161**, 162

commuting 49, **49**, 52, 61, 122
compact cities 64, 65, 121
computer games *see* LogiCity
congestion charging 93
construction industry *see* building codes; building materials
consumer products/patterns 66–68, 108, 117, 229
cool roofs 171–173
Copenhagen Summit/Accord (2009) 6, 9, 106, 223
coping strategies 35, 104, 186; and adaptation 144–146, 178, 179, 231–232; household *41*, 160–161, 171; traditional/informal 150, 160–161, 175–176, 178
Costa Rica 160
Council of European Municipalities and Regions (CEMR) 75
CRAGs (Carbon Rationing Action Groups) 223–224
cycling 127, 128

dairy supply chain 67–68, *68*
Dar es Salaam (Tanzania) 160, 162, 178
data 49–50, 53, 236; access to 51–52, 60–61, 98–100, *101*, 112, 152–153; *see also* knowledge
The Day After Tomorrow (2004 film) 10
deforestation 109
delta regions 28–29
demand management 125–126, 178
demonstration projects 80, 93, 123, 129, 130, 194, 196
Denmark 54
developed countries 29, **73**, 107, 126, 150–151, 154, 168; GHG emissions of 46, 59, 61; governance in 93, 97
developing countries 6, 8–9, 29, 46; carbon finance in 55; GHG emissions of 53, 59, 61–62; transnational networks in **79**, 80–82
Dhaka (Bangladesh) 38, **39**
disaster response/management 150, 164–167, 179
disease *21*, **23**, 37, **39**, 40, *41*
district heating systems 129
Dockside Green (Victoria, Canada) 121

Dongtan eco-city (China) 121
Dortmund (Germany) 60, *60*
drainage systems 32, 34, 36, 127, 155, 160, 178
drought *3*, 6, 10, 20, *21*, 32, *41*, 42, 164, 230
Durban (South Africa) 137, 158–159, **172**, 180, 184, 212

Earthship Brighton (UK) 224
eco-cities 15, 121, 206, 226, 235, 236
economic development 29, 32, 83, 85, 87, 158, 186, 237
economic impact of climate change 24–25, 28
elderly people 40–42, 155, 165
electric vehicles 114–115, 127–128
electricity grid 129–130, **131**
emissions trading schemes **203–205**
emissions/carbon intensity 63, 108, 130
employment 122, 231
Energie Cités 76, 77, 78
energy demand management 125–126
energy efficiency 76, **79**, 86, 114, 118, 122–126, *139*; and climate change experiments 211, 212, 216–217, 218; and demand management 125–126; and retrofitting 123–124; standards/regulations 123
energy infrastructure 32, 128–130; decentralised 129; in Global South 129; low-carbon 129–130; street lighting *58*, **111**, 114, 128–129
energy production 11, 61–62, 63–64, 68, 106, 119
energy sector 85
energy security 77, 129, 130, 208–209
energy use 8–9, 11, 20, *21*, 22, *33*, 45, 56, 212; monitoring **49**, **50–51**, 125–126; reducing *see* energy efficiency; and urban form/density 64–65; 'value-action' gap in 126
energy vulnerability 216
energy-intensive industries 61–62, 63, 127
environmental justice 5, 16
environmental organizations 72
Environmental Protection Agency (EPA, US) 75

Europe 7, 24, 40, 82, 103, 104, 107, 132, 153, 198; adaptation in 179, 184; climate change impacts in *3*; climate-change networks in 75–76
European Covenant of Mayors (CoM) 78, 83
European Union (EU) 84, 136, 184
extreme events/weather *3*, 19, 20, 23, 36, 40

FCPF (Forest Carbon Partnership Facility) 54
financial sector 84
flooding *3*, 6, 20, *21*, **23**, 28–29, 32, *41*, 42, 229, 230; adaptation and 160, **161**, 168, 169, 170–171, 175–176; cost of 24; and poverty 34–35, 36, **39**; social impacts of **39**; variations in vulnerability to 35–38; *see also* sea-level rise
floodplains 32–34, 36, *183*
food security *21*, 22, 37, *41*, *58*, 158, **172**
food supply 52, 67–68, *68*, 121
Forest Carbon Partnership Facility (FCPF) 54
forestry 56, 171; urban 25–26, *58*, 108–109, 153; *see also* tree planting
Forward Chicago initiative 102
fossil fuels 61–62, 63, 106, 108, 115, 119, 130
Frankfurt (Germany) 74, 75
Freiburg (Germany) 179
Friends of the Earth 212
futures, climate-change 4, 9–14, *13*; dystopian 10, 13; utopian 11–13

G8 countries 78
garden cities 11, 206
gender 35, 231
Germany 60, *60*, 74, 75, 76, 106, 123, 128, 129, 179, 193
Ghana **161**, *162*
GHG (greenhouse gas) emissions 4, 5, 45–70; and car ownership 65; and climate 64; confusion in debate over 46; and consumption patterns 66–68; described **2**; domestic 64, 65–66; dynamics/drivers of 63–68; and food supply 67–68, *68*; further reading/

260 • Index

resources on 70; measuring/monitoring *see* GHG emissions accounting; per capita 61–62, **62**, 63, 69; reduction, international finance for 54, 69; rise in levels of 2, 106; sites/processes of 46, 47, 48, 49, 52–53, 55, 107, 232–234; targets/timetables for *see* GHG emissions targets; and urban diversity 59–63, 69; urban emissions of 7–8, *8*, 45–46, 58–63, *60*; and urban form/density 64–65, 69; urban responses to 46–47; and wealthy/poor cities 59, 60–61, 62–63, 66–67

GHG emissions accounting 47–54, 84, 109, 192, 202; boundaries and 52, 53; ClimateCam model 50, **50**, *51*; consumption-based approach to 47, 53, 64, 66–68, 69; and data availability 51–52; data level in 48–50, 52; integrated approach to 53–54; inventories 47, **48**, 50, 51, 52; Local Government Emissions Analysis Protocol 47–50, **48**; production-based approach to 47, 52–53, 60, 61, 64–66, 69, 232–233; and Scope I/II/III emissions 48–49, **49**; and sites/processes of emissions 46, 47, 48, 49, 52–53, 55, 232–234; *see also* carbon finance

GHG Emissions Analysis Protocol, Local Government 47–50, **48**

GHG emissions inventories 47–51, **48**, 53, **79**, 80; further reading/resources on 70; and mitigation policies **111**, 114; and urban boundaries 52

GHG emissions targets 5–6, 74, 76, 83, 118; absolute/relative 108; and mitigation policies 106, 111–112, **113**; zero 86, 87, *88*, *89*

glacial retreat 20–21, **22**

global cities 80, 82, 192

Global Cities Institute (Aecom) 82

Global Cool Cities Alliance 173

Global Report on Human Settlements 2011 (UN-Habitat) 80

governance, climate-change 14–15, 71–105, 135–136, 190–191; adaptive 197; and climate change experiments 195–197, 198, 201–202, 226; diverse range of 71–72, 82; drivers of/barriers to 73, 77, 91, 98–104, *101*; evaluating 82–91; further reading/resources on 105; history of 73, 74–91; institutional factors in 98–102, *101*, 192; international 54, 72; leadership in 86–87, 102; Melbourne case study 83, 84, 85–91, *88–89*, *90*; modes of 91–98, *94–95*, 104, 119, 164, 179, 184; multilevel 72–73, 77, 85, 100, 135, 184; municipal enabling mode 95, 97, 98, 193–194, 209; municipal provision mode 93, *94*; municipal regulation mode 93–97, *94*; municipal self-governing mode 92–93, *94*; municipal voluntarism approach *see* municipal voluntarism; non-state actors in 82, 95, 98; non-state mobilization mode *96*, 98; partnerships in *see* partnerships; and policy entrepreneurs 87, 102, **103**; political factors in 98, *101*, 102–103; public–private provision mode 95–*96*, 98; regional/national campaigns 77–78; resources for 100–102; and scale 71–72, **73**; sectoral 72; self- 193–194, 233; sociotechnical factors in 98, 103–104; 'soft' regulation in 95, 98; strategic urbanism approach 74, 77–82, 83–84, 97–98, 104; and transitional adaptation 148; transnational networks *see* transnational networks

grass-roots/bottom-up approach 11, 50, 142, 184, 196, 215, 220, 225

Great Barrier Reef (Australia) *3*

Green Growth Program (Amman) 55–58

green roofs 25, *89*, 173, 179

Greenworks Philadelphia 156–158

grey water 130, 173, **174**

Halifax (Canada) 153

halocarbons **2**

Hamburg (Germany) 128

Hammarby Sjöstad (Sweden) 130

HCCAS (Headline Climate Change Adaptation Strategy, Durban) 158

health effects of climate change *3*, 18, 20, *21*, 32, *33*, 36, 37, 65, 77, 155; and

adaptation strategies 165; children's 40; mental *21*, 39, *41*
heating/cooling 63, 64, 65, 108, 115–116, 130, 171–173, 176; CHP/district heating systems 129
heatwaves *3*, 19, *21*, 23, 25, 29, 40–42, *41*, 229; adaptation and 153, 154–155, 156–157, 164, 165, **166**, 171; warning systems 165, **166**
Heidelberg (Germany) 74
Ho Chi Minh City (Vietnam) 153
Hong Kong 7, *31*, 82, 121, 125, *139*; 'Power Smart' campaign in 212, *213*
hospitals 122, 123, **124**
households *88*, 115–116, 123, 125–126, 129, 212, 216–217; and adaptation 152, 161, 165, 173; energy demands of 64, 65–66
housing 122, *133*, 155, 224; retrofitting in 123–125, 157, **174**
Houston (US) 61
HSBC 82
hybrid vehicles 127–128
hybridized climate change strategy 215, 216–217
hydro-electric power 21, *21*, *41*, 63
hydrofluorocarbons **2**, **203**
hydrogen-powered vehicles 127–128

ICLEI (Local Governments for Sustainability) 47–50, 52, 53–54, 85, 110, 194, 236; CCP programmes *see* CCP (Cities for Climate Change) programme; and private actors 82
IEA (International Energy Agency) 8, 10–11, 45, 47
India 1, 6, 82, 123, **124**, 180, 206–207; ICLEI/Solar Cities Programme in **79**, 80; poverty in 160; *see also* Mumbai
indigenous peoples 75, 163
Indonesia 82, 120
industry 62
informal settlements 32–34, 51, 52, 120, 160, 168, 170, *181*, 184; resettlement of 178
infrastructure 18, *21*, 29, 32, *33*, 34–35, **39**, *41*, 229; and adaptation 148, 151, 158, 161, 167, 176–179, **177**, 180, *181–183*, 188; and climate-change governance 76, 92, 93, *95*, 103–104; energy 32, 128–130; in Global South 126, 127; green 158, 179; and mitigation 107, 126–130, **131**, 233; and mitigation, drivers/barriers in 132, *133–134*, 138; and post-networked urbanism 130; projects 54; transport *see* transport systems
innovation 9, 11–12, 14, **124**, 187; policy 15, 192, 201–206, **203–205**; social 198–199, 201; *see also* climate change experiments
insulation 103, 155
insurance 24, 84, 169
international community 1, 5–6, 14, 83, 106, 143, 195, 209, 237
International Energy Agency (IEA) 8, 10–11, 45, 47
international governance 54
International Union of Local Authorities 75
IPCC (Intergovernmental Panel on Climate Change) **2**, **19**, 28, 148; Fourth Assessment Report (2007) 1–2, 18, 35
Italy 54, 76, 128

Jakarta (Indonesia) 120
Japan 7, *60*, 135
justice: climate 235–237; environmental/social 5, 15, 16, 82, *95*, 121, 192, 215, 216–217, 223–224, 225, 226, 235

Karachi (Pakistan) 7
Kimbisa, Mayor Adam 142
Kinshasa (DRC) 60, *60*
Kirklees (UK) 74
knowledge 27, 38, 43, 113; access to 72, 78, 87, *95*, *101*, 160–161; *see also* data
Kyoto Protocol (1997) **2**, 5–6, 54, 77, 78, 106, 198

Lagos (Nigeria) 34, 38, **39**, 60, *60*
land erosion/subsidence *21*, 28–29
land-use planning/zoning 25, 91, 93, *94*, 119, 120, 132, 168, 180
Large Cities Climate Leadership Group 117, 156

Latin America *3*, 32–34, 63, 80, 127, 132, 179; CCP programme in 76, 78
LCMP (Low Carbon Manufacturing Programme) 212
LCMP (Low-carbon Office Operation Programme) 212
LCZs (low carbon zones, London) 219–222
LDA (London Development Authority) 122, 129
leadership 86–87, 102, *133*, 136, *182*, 185
least developed countries 7, 8–9
Leicester (UK) 74
lighting 66, 72
lighting, street *58*, **111**, 114, 128–129
livelihoods *21*, 24, 28, 36, 37
living laboratories 196–197
Livingstone, Ken 78
Local Agenda 21 74
Local Government GHG Emissions Analysis Protocol 47–50, **48**
LogiCity **12**
London (UK) 7, 10, *60*, 66, 67, 78, 82, 135; adaptation in 166, 168, 169, 180; climate change experiments in 211; climate risk assessment for 153; Development Authority (LDA) 122, 129; electric vehicles in 128; heat networks in 129; low-carbon policies in 122, 123; Thames estuary project 168; Transition Towns in 218–222
Los Angeles (US) 179
Low Carbon Manufacturing Programme (LCMP) 212
Low Carbon Trust 224
low carbon zones (LCZs, London) 219–222
low-carbon cities 113–117, 121–122
Low-carbon Office Operation Programme (LCMP) 212

Madrid (Spain) 153
mal-adaptation *183*, 186–187
malnutrition 40, *41*
Manchester (UK) **99**, 153, 211
Mannheim (Germany) 76
manufacturing sector **49**
marginalized people 5, 7

Masdar City (UAE) 206
mayors 9–10, 78, 83, 102, 111, 117, 129, 136, 142, 156, 194; *see also* CoM; US Mayors' Agreement
media 1, 6, 10
Melbourne (Australia) 15, 83, 84, 85–91, *88–89*, *90*; CCP programme in 85, 86, *88*; energy efficiency in 123, 129; hybridized climate change strategy in 216–217; leadership in 86–87; NAGA and 86, *88*, *89*; partnerships in 86, *88*, *89*; water recycling in 173; zero emissions target in 86, 87, *88*, *89*
methane (CH_4) **2**, 108, **111**, **203**
Mexico 6, 82, 170, 186
Mexico City (Mexico) **22**, 34, 135–136, 137
Miami (US) 7, **23**, 24, *60*
Middlesborough (UK) 211
migration 10, *41*, 52
'milestones' approach 110, **111**
milk supply chain 67–68, *68*
MillionTrees campaign (New York) 222–223
mitigation 15, 55, 58, 69, 82, 84, 93, 104, 106–141, 156, 163; and adaptation, compared 185–186, 187; and built environment *see* built environment and mitigation; and climate change experiments 190, 192, 198, 226; definition/diverse approaches to 108–109, 112–113, 118–119; drivers/ barriers of 107, 112, 132–138; finance for 136; further reading/resources on 141; and GHG emissions targets 106, 108, 111–112, 113; goods/services excluded form 117; importance of municipalities in 107; and infrastructure systems 107, 126–130, **131**, 179, 198; institutional factors in 132–136, *133*, 140, 192, 233; low take-up of 117; 'milestone' approach to 110, **111**; monitoring 10, 110–112, 113, 117, 143, 232–233; and municipal powers 123, 132–135; and partnerships 116–117; policy, development of 107, 109–118; political factors in *133–134*, 136–137, 140; São Paulo case study 113–117, *115*, *116*; sociotechnical factors in *134*,

138, 140, 233; transnational networks and 110, 234; and urban development 107, 119–122, 198
mixed-use development 56, *57*, 93, 120
mobility *41*, 65, 84, 102, 119, 127, 229
Montreal Protocol **2**
Moreland Solar City project (Melbourne) 216–217
Moscow (Russia) 60, *60*
Mumbai (India) 7, 24, *30*, *31*, 82, 129, 170; adaptation to flooding in 175–176, 178; rainwater harvesting in 173, **174**, *175*
Munich (Germany) 74, 123
municipal authorities 5, 15, 36, 72; and carbon finance 54–55; climate change governance by *see* governance, climate-change; and energy sector 119; GHG emissions accounting by 47–54; in partnerships *see* partnerships; responses to climate change by *8*, 9
municipal voluntarism 74–77, 104; and transnational networks 75–77

NAGA (Northern Alliance for Greenhouse Action, Australia) 86, *88*, *89*
Nagpur (India) **79**
natural gas 63, 106, 129
natural habitats *3*
NDTV-Toyota Greenathon 1
Netherlands 54, 123, **131**, 153, 169
'new catastrophism' 22
New Orleans (US) 38
New York (US) 7, **23**, 24, *30*, *31*, 75, 82, 173; adaptation in 173, 176–178, **177**, 179, 180; GHG emissions of *60*, 61; MillionTrees campaign in 222–223; mitigation policies in 111, 123; urban heat island initiative (NYCRHII) 25–26, 27; water/wastewater management in 176–178, **177**
New Zealand *3*
Newcastle (Australia) 50, **50**, *51*, 125
Newcastle (UK) 74, 76
NGOs (non-governmental organizations) 6, **73**, 98, 150, 163, 185, 196, 217
nitrous oxide (N$_2$O) **2**, 108, **203**
North America 7, 82, 104, 107, 198, 216; climate change impacts in *3*

Northern Alliance for Greenhouse Action *see* NAGA
Nottingham (UK) 218; Declaration on Climate Change 83
NYCRHII (New York City Regional Heat Island Initiative) 25–26, 27

ocean circulation patterns 23–24
OECD (Organisation for Economic Co-operation and Development) 8–9, 46, 127
oil 63, 129; peak- 218
'On the Changing Atmosphere' conference (Toronto, 1988) 74
Ontario (Canada) 178
'Oxford is My World' campaign (UK) 211, **211**
ozone (O$_3$) **2**, 108

Paris (France) 128, 166
partnerships 72, **73**, 82, 86, *88*, *89*, **99**, 102, *181*; and climate change experiments 199–200; and community-based adaptation 162–163; and mitigation policies 116–117, 122, **124**, 129, 132, *133*; public–private provision mode *95–96*, 98
payback periods 100, *133*, 136
PCF (Prototype Carbon Fund) 54
peak-oil challenge 218
Penn State/NCAR Mesoscale Model 25
Pennsylvania (US) 202–206
perfluorocarbons **2**, **203**
Philadelphia (US) 15, 123, 154–158, *157*, 166, 173
Philippines 160, 163
Poland 77
policy entrepreneurs 87, 102, **103**, 184
policy innovation 15, 192, 201–206, **203–205**, 235
political cycle 136–137
pollution *3*; *see also* air quality/pollution
population growth 4, 7, 29, 32, 45, 56
port cities 28–29
Portland (US) 137
Postcards from the Future (art exhibition) 10

poverty 18, 19, 32–35, 36–37, 39–40, **39**, 217; and adaptation 155, 158, 159–163, 167, 231; and GHG emissions 66; *see also* informal settlements
'Power Smart' campaign (Hong Kong) 212, *213*
predicting climate change 22–27, 29, 152–153; models for *see* climate change models; policies and 27; and scale/local context 25, 26–27; and timing/extent of impacts 22–24
private sector *8*, 52, 54, 82, 192, 199–200; *see also* partnerships
PROMISE-Bangladesh pilot project 165–167
Prototype Carbon Fund (PCF) 54
public buildings 114, 122
public space 72, 84
public transport *see* transport systems
PVA (participatory vulnerability analysis) 27

Quito (Ecuador) **22**, 129, 163

rainfall patterns 20, **22**, 23, 29, *41*, 155
rainwater harvesting 130, 173, **174**, 224
recycling 52, *57*, 130; *see also* water conservation/recycling
REDD (reducing emissions from deforestation and degradation) 109
regional governments 72, **73**
renewable energy *58*, 63, 78, **79**, 86, 104, **111**, 211, 224; and 'smart city' projects 129–130, **131**; solar 63, 115–116, *116*, 129–130
resettlement 178
resilience 15, 38, 80, 83; and adaptation 147–148, *149*, 153, 159, **161**, 162, 165, 171, 178, 232; of buildings 171; and climate change experiments 215, 216, 217–223; defined **147**
Resilient Cities conference 80
retrofitting *89*, **111**, 118–119, 123–125, 157; in low-/middle-income economies 125, **174**, 217
Rio de Janiero (Brazil) 125, 129
risks of climate change *3*, 4, 6–8, 18–20, *21*, 43, 229, 231; adaptation and 142,
146, 152, 160, 164–167, 168; assessing nature/level of *see* predicting climate change; cities as cause of 7–8; and tipping point 23–24
Rockerfeller Foundation 80
Rome (Italy) 76, 128
Rotterdam (Netherlands) 123, 153
Russia 7, 60, *60*

St Petersburg (Russia) 7
salinization *21*, *41*, 157
San Francisco (US) 128, 130
sanitation systems 32, 63, 127, 138, 146, 158, 160, 161, 171, 173, 176, 178, 231; *see also* sewage systems
São Paulo (Brazil) 15, 113–117, *115*, *116*, 123, 137
sea-level rise 1, *3*, 10, 230; and adaptation 153, 169; vulnerability to 19, 20, *21*, 23, **23**, 28–29, *30–31*, 36, *41*, 42
seasonal impacts of climate change *33*, 34
Seattle (US) 78, 153–154
self-governance 193–194, 233
self-sufficiency 121, 130, 218
service industry 62, 117, 122
sewage systems 34–35, 155, 157–158, 160; *see also* sanitation systems
Shanghai (China) 7, *30*, *31*, *60*, 61–62, 82; Dongtan eco-city 121
slums *see* informal settlements
'smart city' projects 129–130, **131**
smart meters 129, 218
social capital 40, 42, 146, 150, 151
social innovation 198–199, 201
social justice *see* justice, environmental/social
social networks 40–42, *41*, 222–223, 231
Solar American Cities programme 129–130, 209, **210**
Solar (McEwan) 10
solar power **124**, 129–130; hot-water systems 63, 115–116, *116*, 123
solar thermal ordinances 123
South Africa 62, 82; *see also* Cape Town; Durban
South Asia 78, **79**, 80
South East Asia 78
South Korea 82

Spain 54, 123, 128, 153
Stern Review (2006) 7, 45
Stockholm (Sweden) 76, 123
storms 20, *21*, 23, 24, 29, 34, 36, *41*, 42, **161**, 164, 229; and adaptation 158, **161**, 164, 171; run-off from 32, 153, 158, 173
strategic urbanism 74, 77–82, 83–84, 97–98, 104, 226
street lighting *58*, **111**, 114, 128–129
Stuttgart (Germany) 179
subsidies 93, *94*
sulphur dioxide (SO$_2$) **203**
sulphur hexafluoride **2**, **203**
supermarkets 72, 230
surface treatments 25, 173
sustainable construction 114
sustainable development 9, 57, 74–75, 121, 122, *183*, 236; experiments in 215, 223–224, 235
sustainable energy *see* renewable energy
Sweden 76, 122, 123, 130, 135
Sydney (Australia) 130, 169

T-Zed (Towards Zero Carbon Development, India) 206–207
Tanzanian Federation of the Urban Poor 162
taxation 93, *94*, 132
technological innovation 224, 225, 226, 235
temperature rise 2, *3*, 20, *21*, **22**, 23
Thames Estuary 2100 (TE2100) project 167
Thane (India) 123, **124**
tipping point 23–24
Tokyo (Japan) 7, *60*
Toronto (Canada) **23**, 60, *60*, **62**; adaptation in 165, **166**; climate-change governance in 74, 75
Toronto People's Assembly 223
tourism 20, *41*, 186
traffic congestion 77, 93, 128
Transition Towns movement 12–13, 82, 215, 217–223
transnational municipal networks 75–80, 83, 100, 236; and climate change experiments 192, 194, 201; in Global South 80–82; and mitigation 110, 234; *see also* C40 Cities; CCP; Climate Alliance
transparency **48**
transport systems 9, 32, *33*, **39**, 72, 114–115, 121, 127–128; BRT (bus rapid transit) 114, 115, 127; commuters and 49, **49**, 52, 61, 122; congestion in 77, 93, 128; in developing countries 127; and GHG emissions 46, **49**, 52, *57*, 103, 127; and governance 72; and mitigation 114–115, 119, 127–128, 132, 198; *see also* car ownership/use; mobility
tree planting 25–26, 108–109, 121, **172**, 179, 191, 219; MillionTrees campaign (New York) 222–223
TTB (Transition Town Brixton) 218, 220–222

UHI (urban heat island) effect 25–26, **26**, 32, *41*, 154–155, 165, 173, 179
Umbrella Carbon Facility (UCF) 54
UN-Habitat 53, 59, **73**, 80, 98, **99**, 151, 152, 163, 164, 209
uncertainty 24, 142–143, 180
UNFCCC (United Nations Framework Convention on Climate Change, 1992) 5, 85, 106, 109, 113
United States (US) 24, 32, 84, 153–154, 173, 202–206; adaptation in 153–158, 179; climate change denial in 78; climate-change governance in 78, 103; Environmental Protection Agency (EPA) 75; and GHG emissions 6, 60, *60*, 61; and ICLEI 75, 76; mitigation policies in 103, 123, 129–130, 135; solar power in 129–130, 209, **210**; urban sprawl in 103, 120; 'weatherization' projects in 123; *see also* New York; Philadelphia
urban agriculture *58*, 163, 219
Urban CO$_2$ Reduction Project 75
urban development 9, 20, 32–34, 65, 71, 72, 76, 102–103, 104; and adaptation 160, 167–171, 179, *181–183*; informal settlements and 160, 168, 170, *181*; and mitigation 119–122, 127, 132, *133–134*, 198, 233; and poverty 160, 167

urban diversity 59–63, 69, 71–72, 82, 150–151, 229–230, 237
urban forestry 25–26, *58*, 108–109, 153, 179; *see also* tree planting
urban form/density 11, 64–65, 120–121, 127, *134*, 138
urban metabolism 67
urban planning 9, 112, 120, 135–136, 148, 186, 199, 207; *see also* building codes
urban sprawl 103, 119–120, 127, *133*
Urban Transitions and Climate Change project *see* UTACC
urban vulnerability 9, 14, 18–20, 27, 28–44, 84, 229, 236; and adaptation 144, 146, 160, 188, 230–232; Cape Town case study 36–38; children/elderly people and 39–42, *41*; cities as cause of 7–8, 29–30; defined **19**; energy 216; further reading/resources on 43–44; and geographical location 19, 28–29, *30–31*; and income groups 34–35; and infrastructure systems 32, *33*, 38, **39**; and poverty 32–35, 36–37, 39–40, **39**, 42–43; predicting 23–24, 29; and risk 28, 32–34; as social process 28, **39**, 40–43; and urban development 32–34, 38
urbanization *3*, 7, *8*, 9, 18, 29, 32–34, 35, 69, 104
US Mayors' Agreement 78, 83, 156
USAID (US Agency for International Development) **79**, 165–167
UTACC (Urban Transitions and Climate Change) project 197–200, *199*, *200*
utilities 9, 22, 29, 68, 72, 171

Vajxo (Sweden) 122
Victoria (Canada) 121
Victorian Greenhouse Strategy (Australia) 85–86
Vienna (Austria) 123

warning systems 165, **166**, 170
waste management 9, **49**, **50**, *57*, 68, 93, 119, 127, 158, 160; and renewable energy **111**, 114, 129
wastewater treatment 32, 34
water conservation/recycling *58*, 130, 206, 224
water run-off 32, 34
water security 130, **172**
water services 9, 22, *33*, 34, 36, **39**, 49, 68, 94; and adaptation 158, 170, 175–177, **176**, 179, 197; energy-intensity of 127; innovations in 206, 207–208, *208*; and mitigation 119
water shortages *3*, 10, 20–21, *21*, 32, 34, *41*, 56, *58*; and adaptation 155, 173, **174**; and conflict **22**
wildfires *3*, 37
wind power *58*, 63
women 5
World Bank 54, 55–56, 59, 60, 61, 80, 236
WWF (World Wide Fund for Nature) 67, 212
WWF Earth Hour campaign 1

Yogyakarta (Java) 118–119, 128–129

zero emissions target 86, 87, *88*, *89*